$\mathscr{M}$ICROSCOPIC $\mathscr{E}$XPLORATIONS

A GEMS Festival Teacher's Guide

Grades 4–8

Skills
Observing, Comparing, Measuring, Recording Data, Inferring, Analyzing, Communicating, Manipulating Laboratory Materials and Equipment

Concepts
Lenses, Magnification, Images, Properties, Shape Recognition, Pattern, Similarity, Comparison Testing, Crystals, Dissolving, Evaporation, Aquatic Habitats, Life Cycle, Animal Structures and Behavior, Plant Structures, Research Techniques of Scientists

Themes
Scale, Structure, Patterns of Change, Systems and Interactions, Diversity and Unity

Mathematics Strands
Pattern, Measurement, Function, Geometry

Time
Variable, depending on format: One 90–120 minute session; *or* two or more 45- to 60-minute sessions; *or* ten 10–20 minute single learning station activities

by **Susan Brady** *and* **Carolyn Willard**
with contributions from **Michael Isaacson** *and* **Caroline Schooley**

Sponsored by a contribution from the Microscopy Society of America (MSA)

LHS GEMS

Great Explorations in Math and Science
Lawrence Hall of Science
University of California at Berkeley

Cover Design
Carol Bevilacqua

Illustrations
Lisa Klofkorn

Photographs
Richard Hoyt
Laurence Bradley
Susan Brady
Janet Schwarz

Lawrence Hall of Science, University of California, Berkeley, CA 94720-5200

Chairman: Glenn T. Seaborg
Director: Ian Carmichael

Initial support for the origination and publication of the GEMS series was provided by the A.W. Mellon Foundation and the Carnegie Corporation of New York. Under a grant from the National Science Foundation, GEMS Leader's Workshops have been held across the country. GEMS has also received support from: the McDonnell-Douglas Foundation and the McDonnell-Douglas Employee's Community Fund; the Hewlett Packard Company; the people at Chevron USA; the William K. Holt Foundation; Join Hands, the Health and Safety Educational Alliance; the Microscopy Society of America (MSA); the Shell Oil Company Foundation; and the Crail-Johnson Foundation. GEMS also gratefully acknowledges the contribution of word processing equipment from Apple Computer, Inc. This support does not imply responsibility for statements or views expressed in publications of the GEMS program. For further information on GEMS leadership opportunities, or to receive a catalog and the *GEMS Network News*, please contact GEMS at the address and phone number below. We also welcome letters to the *GEMS Network News*.

Development, testing, and publication of *Microscopic Explorations* were made possible by a contribution from the Microscopy Society of America (MSA). The Great Explorations in Math and Science (GEMS) program and the Lawrence Hall of Science greatly appreciate this support. In addition to the MSA, contributors to the development of these activities include: the Cornell University Materials Science Center, the Glaxo Wellcome Foundation, the Hertz Foundation for the Advancement of Applied Physical Science, the Hewlett Packard Foundation, and the National Science Foundation.

International Standard Book Number: 0-924886-00-5

Visit GEMS
at
www.lhsgems.org

COMMENTS WELCOME !

Great Explorations in Math and Science (GEMS) is an ongoing curriculum development project. GEMS guides are revised periodically, to incorporate teacher comments and new approaches. We welcome your criticisms, suggestions, helpful hints, and any anecdotes about your experience presenting GEMS activities. Your suggestions will be reviewed each time a GEMS guide is revised. Please send your comments to: GEMS Revisions, c/o Lawrence Hall of Science, University of California, Berkeley, CA 94720-5200. The phone number is (510) 642-7771 and the fax number is (510) 643-0309. You can also reach us by e-mail at gems@uclink4.berkeley.edu or visit our web site at www.lhs.berkeley.edu/GEMS.

LHS GEMS

Great Explorations in Math and Science (GEMS) Program

The Lawrence Hall of Science (LHS) is a public science center on the University of California at Berkeley campus. LHS offers a full program of activities for the public, including workshops and classes, exhibits, films, lectures, and special events. LHS is also a center for teacher education and curriculum research and development. Over the years, LHS staff have developed a multitude of activities, assembly programs, classes, and interactive exhibits. These programs have proven to be successful at the Hall and should be useful to schools, other science centers, museums, and community groups. A number of these guided-discovery activities have been published under the Great Explorations in Math and Science (GEMS) title, after an extensive refinement and adaptation process that includes classroom testing of trial versions, modifications to ensure the use of easy-to-obtain materials, with carefully written and edited step-by-step instructions and background information to allow presentation by teachers without special background in mathematics or science.

Staff

Principal Investigator: Glenn T. Seaborg
Director: Jacqueline Barber
Associate Director: Kimi Hosoume
Associate Director/Principal Editor: Lincoln Bergman
Science Curriculum Specialist: Cary Sneider
Mathematics Curriculum Specialist: Jaine Kopp
GEMS Network Director: Carolyn Willard
GEMS Workshop Coordinator: Laura Tucker
Staff Development Specialists: Lynn Barakos, Katharine Barrett, Kevin Beals, Ellen Blinderman, Beatrice Boffen, Gigi Dornfest, John Erickson, Stan Fukunaga, Philip Gonsalves, Linda Lipner, Karen Ostlund, Debra Sutter
Financial Assistant: Alice Olivier
Distribution Coordinator: Karen Milligan

Workshop Administrator: Terry Cort
Materials Manager: Vivian Tong
Distribution Representative: Felicia Roston
Shipping Assistant: Jodi Harskamp
GEMS Marketing and Promotion Director: Gerri Ginsburg
Marketing Representative: Matthew Osborn
Senior Editor: Carl Babcock
Editor: Florence Stone
Principal Publications Coordinator: Kay Fairwell
Art Director: Lisa Haderlie Baker
Senior Artist: Lisa Klofkorn
Designers: Carol Bevilacqua, Rose Craig
Staff Assistants: Chrissy Cano, Larry Gates, Trina Huynh, Chastity Pérez, Dorian Traube

Contributing Authors

Jacqueline Barber
Katharine Barrett
Kevin Beals
Lincoln Bergman
Susan Brady
Beverly Braxton
Kevin Cuff
Linda De Lucchi
Gigi Dornfest

Jean Echols
John Erickson
Philip Gonsalves
Jan M. Goodman
Alan Gould
Catherine Halversen
Kimi Hosoume
Susan Jagoda
Jaine Kopp

Linda Lipner
Larry Malone
Cary I. Sneider
Craig Strang
Debra Sutter
Herbert Thier
Jennifer Meux White
Carolyn Willard

Cover Photo Credits

Front Cover: photograph of **student and microscope** by Richard Hoyt; **spider** from Joseph P. Neilly, Senior Microscopist, Abbott Laboratories, Abbott Park, Illinois; **currency** by Caroline Schooley; **sand** from Helen Adams, Academically Gifted Resource Teacher, Penny Road Elementary School, Raleigh, North Carolina (via Dr. John Russ). The **background** is derived from an SEM sand montage by Tina Carvalho, University of Hawaii.

Back Cover (from left top): the desmid *Micrasterias* (an example of green algae) and the **dime**, both from Professor Nina Allen, Department of Botany, North Carolina State University, Raleigh, North Carolina; **camphor** from Frank S. Karl, Microview Consulting, Akron, Ohio; the American woman **stamp** by Caroline Schooley; photo of **synthetic knit textile** by Marc Van Hove, 6th prize, 1992 Nikon International Small World Competition; photo of *Daphnia pulex* by Dr. Stephen Lowry, 7th prize, Nikon International Small World Competition (for more information on Small World, phone 516-547-8569); **Vitamin C** by Dennis Kunkel, Microvision, Kailua, Hawaii.

Reviewers

We would like to thank the following educators who reviewed, tested, or coordinated the reviewing of *Microscopic Explorations*. Their critical comments and recommendations, based on classroom and schoolwide presentation of these activities nationwide, contributed significantly to this GEMS publication. Their participation in this review process does not necessarily imply endorsement of the GEMS program or responsibility for statements or views expressed. Their role is an invaluable one; feedback is carefully recorded and integrated as appropriate into the publications. **THANK YOU!**

CALIFORNIA

Albany Middle School, Albany
Kay Sorg *

Canyon Middle School, Castro Valley
David Horsley
Kathleen Lloyd
Scott Malfatti *
Mike Wescott

Fairmont Elementary School, El Cerrito
Nanci Buckingham
Carrie Cook *
David Ko
Laura Peck

Hillview Jr. High School, Pittsburg
Lynette Cardinale
Cheryl McInerney
Elizabeth Ramsey
Guillermo Trejo-Mejia *

King Jr. High School, Berkeley
Karen Hansen
Yvette McCullough
Marie Schumacher
Beth Sonnenberg *
Phoebe Tanner

Tolenas Elementary School, Fairfield
Kathy LaRocco *
Gail Patmon
Darrel Samuels
Tim Smith

CONNECTICUT

John B. Sliney School, Branford
Ginny Baltay
Nicole Binehardt *
Ginger Dendas
Steve Naracci
Marilyn Odell

ILLINOIS

Beardsley Middle School, Crystal Lake
Frances Hicks
Ann Minn

Hawthorne Jr. High–North Campus, Vernon Hills
Sue Berg *
Dave Brown
Joe Omiatek
Anne Reichel *
Anne Rizzo

North Middle School, Crystal Lake
Laurel Hochstetler
Beth Richards *

MINNESOTA

Galtier Math, Science Technology Magnet, St. Paul
Ann Anderson
Aaron Benner
Cynthia Berger
Susan Brazel
Steve Hawkinson *
Candace Jensen
Michael Thole

MISSOURI

Blades Elementary, Melville
Renee Bennerote

NEBRASKA

Kahoa Elementary School, Lincoln
Jeff Brehm
Diane Brown
Mary Buicham
Mike Fangman *
Lance Hall

NORTH CAROLINA

Frank Porter Graham Elementary, Chapel Hill
Gwen Atwater
Rebecca Bennett
Cammie Brantley
Susan Carter
Deanna TeBockhorst *

University of North Carolina at Chapel Hill
Pat Bowers *

OREGON

Phoenix Elementary School, Phoenix
Charlie Bauer *
Erin Mahanay
Kathy McMurtrey
Marlys Meulmans

* Trial test coordinators

The following people were involved in the collaboration that led to the creation of the prototype program on which this guide is based, through Project MICRO of the Microscopy Society of America (MSA).

Bay Area/Northern California

Mei Lie Wong, University of California, San Francisco, Coordinator for the Northern California Society for Microscopy (NCSM)

Andy Chamberlin, California Academy of Sciences

Linda Wraxall, California State Department of Justice

Nancy Smith, California State University, Hayward

Carol Leitch, Harding School, El Cerrito

Meg Hudson, Longfellow School, Berkeley

Betty Merritt, Longfellow School, Berkeley

Catherine Lynch, Malcolm X School, Berkeley

May McKoon, MH Chow and Associates

Steven Shaffer, MicroDataware

R'Sue Caron-Popowich, Parent

Bernard Thomas, RJ Lee Group

Roberto Bonilla,
San Francisco Unified
School District

Jane Gerughty,
San Francisco Unified
School District

Erla Hackett,
San Francisco Unified
School District

Jean Regan, San Francisco
Unified School District

Kristen Sorensen,
San Francisco Unified
School District

Irene Uesato, San Francisco Unified School
District

Ron Breland, St. Paul's
School, Oakland

Cathy Hackler, St. Paul's
School, Oakland

Susan Porter, St. Paul's
School, Oakland

Ann Glimme, Tara Hills
School

Wilfred Bentham, University of California at
Berkeley

Alma Raymond, University of California at
Berkeley

Steven Ruzin, University
of California at Berkeley

Virginetta Cannon,
University of California,
San Francisco

Liesl Chatman, Science
and Health Education
Partnership (SEP)/UCSF

Barbara Plowman,
University of the Pacific
Dental School

Chicago, Illinois

Scientist Partners

Kevin Cronyn, Hitachi
Scientific Instruments

John Lloyd, Northern
Illinois University,
Coordinator for the
Midwest Microscopy and
Microanalysis Society

Joe Neilly, Abbott
Laboratories

Teacher volunteers were
also part of the GEMS trial
testing process, and their
names appear on that list,
at the two schools in
Crystal Lake, Illinois.

Ithaca, New York

Kathleen Hunt,
Sciencenter, Coordinator

Teachers

Rose Bernstein, Vestal
Middle School, Vestal,
New York

Michelle Boyer-White,
West Middle School,
Binghampton, New York

Greg Clark, Waverly
Middle School, Waverly,
New York

Dan Falcone, Dewitt
Middle School, Ithaca,
New York

Diane Gerhart, Candor
Elementary, Candor, New
York

Ronald Hoodak , Elm
Street School, Waverly,
New York

Sally Horak, Cortland
JSHS, Cortland, New York

Barbara Jackson, Corning
Free Academy, Corning,
New York

Cheryl Jones, Corning-
Painted Post School,
Corning, New York

Ray Kaschalk, Jr., Thomas
Edison JSHS, Elmira
Heights, New York

Kathryn Mack, Corning-
Painted Post School,
Corning, New York

Michael Midey, Romulus
Central Schools, Romulus,
New York

Pamela Monk, Ithaca City
Schools, Ithaca, New York

John Morris, Ithaca City
Schools, Ithaca, New York

James Murphy, Watkins
Glen Middle School,
Watkins Glen, New York

Ramona Steigerwald,
Horseheads Middle
School, Horseheads,
New York

Iva Jean Tennant, West
Middle School,
Binghampton, New York

Barbara Wood, Hendy
Avenue School, Elmira,
New York

Scientist Partners

Tom Barbieri,
Cornell University

Chris Bender,
Cornell University

Eileen Fanning,
Corning, Inc.

Melanie Fewings,
Cornell University

John Hallas,
Cornell University

Tanya Jacobs,
Corning, Inc.

Gerald Janauer,
Cornell University

Aaron Judy,
Cornell University

Jens Knoblock,
Cornell University

Andone Lavery,
Cornell University

John Morris,
Cornell University

Theresa Newton,
Cornell University

Jennifer Nichols,
Cornell University

Jeff Potoff,
Cornell University

Paul Pyenta,
Cornell University

Rolf Ragnarsson,
Cornell University

Jinandra Ranka,
Cornell University

Linda Rosenband,
Cornell University

Tiffany Smith,
Cornell University

Irene Tones, Corning, Inc.

Steve Turner,
Cornell University

Victoria Williams,
Cornell University

Portia Yarborough,
Cornell University

Geoffrey Zassenhaus,
Cornell University

St. Paul, Minnesota

Tina Schwach, University
of Minnesota, Coordinator
for the Minnesota Micros-
copy Society

Teachers

Mary Ramoser, Battle
Creek Magnet

Maureen Tauer, Battle
Creek Magnet

Lois London, Como Park
Elementary

Eileen Cotter, Highland
Park Elementary

Ann Scrank, Jackson
Elementary

Shawn Zimmerman,
Monroe Community

Beth Burns, Sheridan
Elementary

Vicki Holst, Sheridan
Elementary

Etta Bernstein, Talmud
Torah of St. Paul

Chris Page, Talmud Torah
of St. Paul

Scientist Partners

Dwight Erickson, 3M

Linda Krier, 3M

Pam Ligget-Olson, 3M

Devora Molitar, 3M

Ev Osten, 3M

Jeff Payne, 3M

David Gohl, Ecolab

Al Edgar, Guidant-CPI

Ivan Lubago, Guidant-CPI

Kerstin Halverson, HCMC

Rodney Rappe, Imation

Jerry Tangen, JEOL

Karen Runyon, Minneapo-
lis Police Department

Joy Frestedt, St. Jude
Medical

John Basgen, University of
Minnesota

Muriel Gavin, University
of Minnesota

Stuart McKernan, Univer-
sity of Minnesota

Foreword

by Dr. Bruce Alberts
President, National Academy of Sciences
Chair, National Research Council

It is a pleasure for me to be able to introduce this book, sponsored by the Microscopy Society of America (MSA), a professional organization of research scientists, in collaboration with the Great Explorations in Math and Science (GEMS) program of the Lawrence Hall of Science.

The National Academy of Sciences was founded in 1863 as a private organization with a charter that requires us to recruit volunteers to advise the U.S. government on matters of science and technology. Today, our operating arm—the National Research Council—oversees the production of nearly one such advisory report every working day. Topics range from analyses of the potential health effects of electromagnetic fields to the safe disposition of the plutonium being removed from dismantled nuclear weapons.

In 1996, we released the *National Science Education Standards,* a roadmap to guide the improvements in science education that will be needed to prepare all of our citizens for the 21st century. This document, like all of our reports, is freely available on the worldwide web at www.nas.edu. Central to the *Standards* is an emphasis on science as inquiry. New emphasis is to be placed on exposing students to activities that investigate and analyze science questions, and on having them use the results of experiments to generate explanations and make scientific arguments.

Experience shows that teachers are much better able to teach science as inquiry, if they have experienced inquiry themselves. In part for this reason, the *Standards* call for making quality science education in our schools the responsibility not only of teachers, but also of many community members who have not previously been involved—including large numbers of volunteer scientists and engineers from both the public and private sectors.

Microscopic Explorations is an outstanding example of curricula being prepared for the schools through a collaboration between volunteer scientists and professional science educators. But it also represents an important device for catalyzing the effective participation of scientists throughout our nation with the teachers in their local schools. Intended to interest and involve all students in their middle school years, it is designed to harness the talents of an important scientific society as a resource in educational reform. Caroline Schooley of the MSA and her many collaborators are to be congratulated for developing this innovative project. But for them, and for the millions of scientists and engineers across the country, the work has only begun. The many scientists who read this guide should carry on what they have so skillfully begun—by volunteering to help teachers use materials such as these to bring the excitement of science to all of their classrooms.

Acknowledgments

The authors and the LHS GEMS program must first of all acknowledge the generous support of the Microscopy Society of America (MSA) in making this guide possible. **Caroline Schooley**, the Educational Outreach Coordinator of the MSA, has been truly inspirational and instrumental in the origination and development of this guide, and in obtaining the material support necessary for its testing and publication. Prior to the development of this GEMS guide, Caroline and Dr. Susan Brady, of the Lawrence Hall of Science developed prototype festival programs for teachers and scientists, including workshop presentations in Ithaca and Minneapolis-St. Paul. Caroline has served as the primary liaison for MSA with the authors and with other GEMS and Lawrence Hall of Science staff. She has enthusiastically and energetically coordinated the efforts of MSA local societies nationwide toward inquiry-based educational efforts, of which this GEMS guide is an important part. MSA local affiliates were enlisted to assist as this guide was developed, and to continue to serve during trial testing and after publication as ongoing resources for microscopes, volunteers, and classroom guests. Caroline wrote the "Selecting School Microscopes" section of this guide, developed extensive resource lists which we have adapted for this guide, and created the drawing of the glass half marble on a leaf on page 27. Thanks Caroline!

Special thanks to Dr. Michael Isaacson, formerly of Cornell, now UC Santa Cruz—a Past President of MSA for his major role in securing funding for this guide, and for his work on the "Special Section on Optics." Thanks as well to Sergeant Allen Yuen of the Crime Scene Unit Berkeley Police Department and Kevin Beals of the Lawrence Hall of Science for their help with fingerprint ridge details. Tara Arnold provided much information on textiles. GEMS Editor Florence Stone contributed to "Behind the Scenes," "Resources," and "Literature Connections," and GEMS Associate Director Lincoln Bergman wrote "Ten Odes to the Microscope."

Our special thanks also to trial test teacher Cheryl McInerney of Hillview Junior High School in Pittsburg, California, for her suggestions that led to great improvements in the format of the student recording booklet. The Discovery Quilt idea is adapted from a similar activity that we first encountered as part of the Family Science program. The water drop magnifier idea appears in various forms in many activity collections and curricula. This particular variation was inspired by a similar activity used in the Family Science program.

The originators of this project wish to make a special commemorative dedication of this book to Wilfred Bentham, one of the first volunteer scientists to work with us, who was tragically killed in a random drive-by shooting during the initial trial testing of these activities.

$\mathcal{F}$or the past several years, the Lawrence Hall of Science has presented Microworld Festivals as large group events. During these festivals, participants move from one station to another, investigating a variety of small things made larger and infinitely more fascinating by magnification. This teacher's guide adapts the Microworld Festival for the classroom, whole school events, or for family or community events.

GEMS festival guides differ from other GEMS guides in that they present students with a variety of different challenges at learning stations set up around the room, rather than a more structured presentation. **A learning station is a table, cluster of desks, or countertop set up with equipment or materials designed to encourage students to make their own discoveries.** The class participates in exploration and investigation in an informal, student-centered way.

Microscopic Explorations is one of several GEMS festival guides. We highly recommend the others, including *Bubble Festival, Build It! Festival, Mystery Festival,* and *Math Around the World.*

Two important things to ask yourself before deciding to present a Microscopic Explorations Festival are:

- *How can I get microscopes, table lamps, and extension cords?*

- *How can I get adult volunteers?*

The number of microscopes and adult volunteers you'll need will vary according to the learning station format you choose, and how many students will participate at one time. For more information on options for festival formats, volunteers, microscopes and related materials, please read "Planning Your Microscopic Explorations Festival" on page 7.

Contents

"*The gifts of microscopes to our understanding of cells and organisms is so profound that one has to ask: What are the gifts of the microscopist? Here is my opinion. The gift of the great microscopist is the ability to THINK WITH THE EYES AND SEE WITH THE BRAIN. Deep revelations into the nature of living things continue to travel on beams of light.*"

Daniel Mazia
U.C. Berkeley cell biologist, 1996

Introduction

Using Microscopes to Draw Students into the World of Science

People love to look at things up close. The *Microscopic Explorations* festival is designed to take advantage of that natural interest and to give students the opportunity to use microscopes and hand lenses to perform a variety of investigations in the life, earth, and physical sciences.

Microscopic Explorations provides rich experiences through which students discover the power of the microscope as a tool for inquiry. The role of microscopes in human inquiry has been demonstrated countless times in many different fields and is reaffirmed in laboratories all around the world each and every day!

In this unit, students do learn about microscopes and gain confidence in their use. It has been our consistent experience that *Microscopic Explorations* festivals effectively defuse anxiety about using microscopes, and replace it with enthusiasm and appreciation for this great tool. That is important, but it is not the primary educational goal of these activities. In harmony with the *National Science Education Standards* and the LHS GEMS approach, this guide is designed to open up the vast realm of scientific inquiry to students. The *National Standards* emphasize that "Inquiry into authentic questions generated from student experiences is the central strategy for teaching science." Activities such as those in this guide provide initial experiences for students in both the content and process of science. Grounded in these activities, students can go on to pursue their own questions in greater depth.

In these activities, students learn how to operate a microscope, to place a slide on the stage, and to adjust the light, but—more importantly—they observe the microscope's usefulness as a tool for scientific investigation.

As one leading microscopist put it, these activities can help students "think with their eyes," that is, learn to observe the world around them critically, with scientific curiosity and skepticism. In time, building upon the understandings and skills gained through such introductory experiences, students can attain a much higher level of scientific literacy—designing their own investigations and experiments, while utilizing the amazing technologies, including microscopes, that human ingenuity has devised.

National Standards Meet Microscopes

In the *National Science Education Standards*, the "Science as Inquiry" standard includes **both** the "Abilities Necessary to Do Scientific Inquiry" and "Understandings About Scientific Inquiry." Up through Grade 4 the *Standards* expect that students become able to "employ simple equipment and tools to gather data and extend the senses." Included in such tools are "magnifiers to observe objects and organisms; and microscopes to observe the finer details of plants, animals, rocks, and other materials." The accompanying understanding about inquiry is that such instruments can provide more information than can be obtained using only one's senses.

In middle school, these inquiry abilities and understandings are raised to a higher level. Students are expected to "use appropriate tools and techniques to gather, analyze, and interpret data," and to understand that "technology can be used to gather data and allows scientists to analyze and quantify results of investigations." The *Standards* also state, "Students in grades 5–8 also have the fine motor skills to work with a light microscope and can interpret accurately what they see, enhancing their introduction to cells and microorganisms and establishing a foundation for developing understanding of molecular biology at the high school level." In addition, students deepen their understanding of science and technology, learning that "Technology is essential to science, because it provides instruments and techniques that enable observations of objects and phenomena that are otherwise unobservable due to factors such as quantity, distance, location, size, and speed." Wherever your students may be on these continuums of ability and understanding, *Microscopic Explorations* will allow them to grow and learn, as they begin to appreciate and harness the "power" of microscopic inquiry.

Placing the activities within this larger context of scientific inquiry can also help allay concerns some teachers may have about not having a complete grasp of the many different fields of science reflected in these activities. The object is to observe, investigate, find out more, and then come up with even more questions. Teachers learning along with their students provide a wonderful role model of lifelong learning. In addition, we have provided concise background information to assist with initial student questions and many resources should you and your students wish to learn more.

The Festival Format

After a short classroom introduction to magnifying lenses and microscopes, the students "learn by doing" in the festival itself. The *Microscopic Explorations* festival consists of ten learning station activities and one summary station. Please see "Overview of the Festival Learning Stations" and the other sections that immediately follow this introduction for more detail. The stations are designed to represent a wide spectrum of scientific investigation and to enliven student curiosity in many different fields of science. This interdisciplinary richness reflects the recommendations of *National Science Education Standards*, the AAAS *Benchmarks*, and other current frameworks in science, technology, and mathematics education.

As the students learn about the specific topics at the stations, they also exercise the important science and math skills of manipulating laboratory materials and equipment, observing, comparing, making inferences, measuring, recording and analyzing data, and communicating with one another. Students expand their understanding of the way microscopic and related techniques contribute to all of these diverse fields, increase their comprehension of modern science and research, and their appreciation for the natural world.

There are many benefits to a festival format. It allows for open-ended free exploration, fosters a spirit of cooperative work, allows students to learn at their own level and pace, and is an exciting whole class or even whole school activity that has everyone talking about science.

Although there is a great deal of preparation required, teachers who have presented these activities are in strong agreement that the impact is well worth the effort. In fact, one of the great advantages of the learning station format is a practical one. For a whole class to study Kitchen Powders at once, you would need to prepare over 100 slides of powders rather than the 18 used in rotation at the learning station. It is relatively easy to gather the few specimens (fabric samples, insects, etc.) needed for one learning station. And one learning station can serve not only a whole class, but a whole school!

The festival format is also perfect for involving outside volunteers such as parents or community members who are interested in assisting and/or participating in educational activities. Each station will benefit from an adult leader to help the students pursue their investigations. Older students may also play this role.

Organization of the Guide

In the first section of this guide, you will find detailed instructions to enable you to put on your own *Microscopic Explorations* Festival, with a variety of presentation options. We have provided an overview of the learning station activities, a section on planning your festival and recruiting volunteers, a step-by-step timeline for preparing for and presenting a festival at your school, and a presentation checklist to help keep things on track.

The next and main section of the guide provides concise instructions for setting up each of the festival stations. For each learning station, there is an overview, materials list, information about how to set up the station, and ideas for going further. In addition, for each station there are station signs with instructions and questions for the students, as well as a Volunteer Sheet to help them assist students in making their own discoveries. For a better understanding of what will happen at learning stations, we recommend that teachers read the Volunteer Sheets as well as the general write-up for each activity. The station signs, volunteer information, and student recording notebook are all intended to be duplicated and permission is not required when they are copied for classroom or teacher's workshop use.

In addition to the Discovery Quilt summary station, the guide includes several other suggestions for providing closure to the festival. As summarized below, explorations at the stations can lead naturally to more in-depth class, group, or individual investigations.

We have also included a wide-ranging "Resources" section that includes other curricula that pursue scientific topics related to microscopy and optics, as well as numerous books, CD-ROMs, videotapes, and web sites for further exploration and investigation. Appropriate literature connections and assessment suggestions are included, as in all GEMS guides.

Custom Video by Warren Hatch

A special custom video by a teacher named Warren Hatch has been specifically designed to match and extend the ten stations in this GEMS guide. The Microscopy Society of America (MSA) asked Hatch, who has made many excellent videos, to produce the video. It is entitled "Video for Microscopic Explorations," is 30 minutes long, and is available from Warren Hatch. Please see the "Resources" section for ordering information on this video, other videos by Hatch, and a number of other resources that could be used in beneficial connection with Microscopic Explorations.

Making Multiple Connections

The learning station format by no means precludes more in-depth study of what is introduced at each station—in fact it can be a great way to introduce more intensive investigations. Student questions will help lead them along the path of further inquiry.

The *National Science Education Standards* strongly recommend that students should have the opportunity to engage in several "full investigations" during the course of the school year. *Microscopic Explorations* could serve as a wonderful launching pad for these investigations.

Inevitably, you and your students will see numerous correlations between each learning station activity and many other parts of your current instructional program. Many of the learning station activities make excellent introductions to other GEMS units, including *More Than Magnifiers, Color Analyzers, Fingerprinting, Mapping Fish Habitats, On Sandy Shores, Stories in Stone, Terrarium Habitats, Schoolyard Ecology,* and *Aquatic Habitats.* For example, you might want your students to learn about fingerprint classification with the GEMS *Fingerprinting* unit in preparation for the *Microscopic Explorations* Fingerprint Ridges station, in which they focus on single details of fingerprint ridges. The *More Than Magnifiers* activities help students delve more deeply into lenses and optics. The GEMS *Color Analyzers* guide constructs student learning about light and color, deepening student understanding of optics, as well as making an intriguing connection to the Dots and Dollars station. Or, after your students are entranced by their microscopic view of the crystalline structure of salt, you might go on to teach the GEMS earth science unit *Stories in Stone,* and expand student understanding of crystal formation and its part in the rock cycle. Or you might tie in crystalline shapes with mathematics using the *Build It! Festival* geometry activities. There are many other good ways, in other GEMS units and in many of the other excellent inquiry-based curricula now available nationally, to extend and deepen the learning in specific content areas. Please see the "Going Further" sections in each activity for our recommendations of GEMS guides and other activities that make strong learning connections.

The *Microscopic Explorations* festival can stand alone as a great exploratory experience that builds the confidence and enthusiasm students need to pursue a deeper understanding of science, math, and microscopy more formally later on. It can also serve as an immediate jumping off point for deeper investigation in any one of the fields of inquiry opened by a particular station or stations. Your own curricular emphases and the investigative interests of your students are likely to open up whole new worlds of inquiry!

Overview of the Festival Learning Stations

Please note: Students use microscopes at all learning stations, except the first two.

1. Up Close

Students construct their own water drop magnifiers, and use water and a variety of transparent objects, including vials and marbles, to further investigate and understand the properties of magnifying lenses introduced in the preliminary whole-class session.

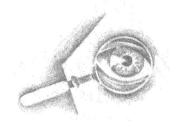

2. Fingerprint Ridges

Using pencils and a piece of tape, students "lift" impressions of their own fingerprints. They use hand lenses to find some of the details of ridge patterns that are used by experts to identify fingerprints. Using their newfound expertise, students carefully examine ridge details of two very similar prints to see whether or not they're identical.

3. Dots and Dollars

Students examine a range of imaging and printing techniques to determine how images are formed. They investigate how dots of only a few colors can be combined in different combinations to give the appearance of a range of other colors.

4. Fabrics

Students observe a variety of fabrics, using hand lenses and microscopes. They compare the appearances and try to determine how each is made. They then select one fabric to examine closely.

5. Salts (*Note:* This learning station includes two activities.)

In the first activity, students observe and compare crystals of various salts and use microscopes to identify which was involved in a make-believe highway spill. In the second activity, they make a crystal suncatcher by coloring a clear plastic lid with crayons and then adding a small amount of a solution of Epsom salts. As the solution evaporates, students watch crystals form on the surface of the lid.

6. Sand

Students compare sand samples from several geographic locations based on the color, size, and shape of the sand grains. They then locate and mark the source of their sand on a map.

7. Kitchen Powders

Students study seven white powders they might find in their own kitchens. They are challenged to observe the properties of the grains of each powder, and to carefully describe and contrast two of the powders. Then, they attempt to identify a mystery powder and a mystery mixture of two powders.

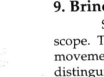

8. Small Creatures

Students enjoy being able to study the fascinating structures of dried insects, spiders, isopods, and more in this high-interest learning station. Focusing on familiar small creatures is one of the time-honored best ways to gain an appreciation of the power of microscopes.

9. Brine Shrimp

Students observe live brine shrimp using a hand lens and microscope. They compare the overall appearance, specific structures, and movement patterns of adults, larvae, and eggs. They learn to identify and distinguish between male and female adult brine shrimp. They also use their math skills to estimate the number of eggs on the slide.

10. Pond Life (*Note:* **This learning station includes two activities.**)

In the first activity, students compare the organisms found in a pond. In the second activity, they use their observations and a variety of art materials to construct a model of a plant or animal they saw with the microscope. These creatures are added to a class mural and compared to other plants and animals observed by classmates.

Discovery Quilt

After they have participated in the festival, students post their findings, questions, and reactions on a class chart. As they think of what to add, they are reminded of what they have seen and learned during the festival, as well as what questions they would like to explore further. The Discovery Quilt is only one of many ways to provide closure to the festival.

Planning Your Microscopic Explorations Festival

Most teachers begin this unit with a whole-class, teacher-led introduction (page 27), and end with one of the whole-class options for reflection (page 99 or 102). Between the introductory and concluding sessions is the diverse festival of learning station activities in which small groups of students "focus" independently on various investigations. There are many different formats in which students can encounter the ten learning station activities that are the heart of this unit.

The number of microscopes, volunteers, space, and time required for your festival will vary with the learning station format you choose, as will the number of students served. In the pages that follow, you'll find a range of possible formats, including: a large room with all stations set up at once; various ideas for offering a few stations at a time; and only one station set up at a time as a learning center in the back of a classroom. Depending on your curricular goals, the age and abilities of your students, scheduling constraints, and possible limitations on number of microscopes and volunteers, you can choose one of the formats outlined below, or use those ideas to design your own format. Whatever format you choose, here are some general answers to the initial questions you may have:

Spanish language translations of the station signs and student sheets in this guide are available from the GEMS Program at Lawrence Hall of Science.

1. How many students per learning station? We recommend planning for four to six students per learning station. The "What You Need" lists for each station accommodate that number of students. However, you can have as many as eight students per station by augmenting the materials at the station. With larger groups it is essential to have tables that are at least six feet by three feet.

2. How many total students? It is easy to expand the ten stations to 12. Two of the learning stations, Salts and Pond Life, each have an accompanying activity in which the students create something related to their discoveries at the station. These two activities (Crystal Suncatcher and the Pond Life Mural) don't require microscopes. They can be set up at separate tables or as part of enlarged Salts or Pond Life stations. This means, for example, that with four to six students per 12 tables, you can accommodate 48–72 students at once in a large group festival. The Discovery Quilt summary station can also be used as an extra station.

Please read over the descriptions of the two additional activities in Pond Life and Salts to determine whether you will include them. While most teachers use the Pond Life Mural, it may be a bit elementary for some (but not all!) 7th and 8th graders. As for the Salts learning station's Crystal Suncatcher activity, it is very popular with students, but does require obtaining lots of clear, plastic lids and a hot pot. The Discovery Quilt activity provides students a way to record their thoughts and questions about their learning station experiences. It can be offered either during the festival or as a concluding activity after the festival. The Discovery Quilt option also may seem elementary for some older students and other options are suggested.

3. How many microscopes? Eight of the festival stations require at least two microscopes each, so you should plan to have at least 16 microscopes. (Stations 1 and 2 don't require microscopes.) If you can't gather 16 microscopes, one format described below allows you to get by with eight or 12. On the other hand, if you can supply more than 16 microscopes, having as many as one per student can be great. Generally, most teachers advise against using inexpensive microscopes with inexperienced students, because they can be frustrating. However, there are some exceptions to this rule. Be sure to read the more detailed information about selecting microscopes on pages 115–119.

4. How many volunteers? The number of adult volunteers needed varies quite a bit, depending on the format you use. For a festival with all the stations at once, you'll need 12 volunteers. For some classroom formats, you can get by with one to six volunteers at a time, but volunteers will be needed on more than one day.

5. Should there be timed rotations to learning stations? Flexibility and a sense of free exploration are highly desirable in the learning station format, and it's fine if not everyone gets to every station. In a large-group festival for families, for instance, if you want participants to explore learning stations at their own pace, you probably won't want to have a schedule of timed rotations. For students, however, many teachers feel that setting a certain amount of time at each station and a system for rotating through the stations is important for classroom management.

With the timed rotation system, students must stay at a station until a signal that it's time to change. Most teachers let students know ahead of time how much time they'll have at each station, assign groups of students to their first station, and make clear which groups will move to which station. Students needn't visit the stations in numerical order. The signal can be flicking the lights, ringing a bell, or whatever you usually use. Giving a two-minute warning before the signal to change to a new station is also important. This way, students can finish observations and/or recording, and tidy up the station for the next group.

If you decide on timed rotations, **remember that there is a relationship between the amount of time students spend at a station and the depth of their learning.** Some formats below provide students about 10 minutes at each learning station. This type of introductory exposure to a variety of investigations is ideal for allowing large numbers of students to become confident, enthusiastic users of microscopes and magnifying lenses. Short stays at each station will allow most students to enjoy some discoveries, consider the questions on the station sign, and record their observations. However, if your goal is for students to make more thorough observations and have time for more thoughtful, detailed recording, choose one of the formats that provides at least 20 minutes per station. In some of the classroom formats below, you can provide longer stays at stations by extending the length of the festival over more than one day.

One teacher who opted for a large group festival with 10 minutes per station told us "There is usually very little exposure to microscopes in the elementary grades, and even in later schooling (and life) for many people. Many students and adults associate microscopes with a distant, challenging realm inhabited only by well-trained specialists...I overheard a fourth-grader who was entering the cafeteria for our festival say to a friend, 'I'm scared!' But very soon, this student, along with all the rest, was enthusiastically using microscopes—building skills and confidence."

Another teacher, who provided students 20 minutes per station in a classroom format, told us that student accountability and careful recording are important goals for her: "I like this type of learning station approach, but I want students to really think about their scientific observations."

A Range of Learning Station Formats

Offering All Learning Station Activities at Once

— as a school program for multiple classes. Many teachers choose to present *Microscopic Explorations* to one or two classes at a time, throughout one or more school days. A large classroom, a multi-purpose room, or other large space is most appropriate for this format.

— as an evening program for parents and community members. To encourage more interaction between school and community and promote scientific inquiry, your *Microscopic Explorations* Festival can take the form of a parent or community open house or a family science evening at a school or community center.

— as a combination school day and evening event. You might consider leaving the festival stations up after the school day event, and having selected students help the adult volunteers guide families through the activities that same evening.

With ten different investigations, this big festival format is especially exciting. Working together to learn can be a very effective community-building experience. If you would like to serve 48–72 students at a time, with a one-time set-up of the festival, this may be the best format for you. By offering more than one session during the day, you can increase the number of students served. Allow at least 90–120 minutes per session. To present all the learning stations at once, you'll need at least 16 microscopes, 12 adult volunteers, a large room with 12 or more tables, and plenty of electrical outlets.

Note: You can involve more than 72 students at once by making multiple learning stations of some of the activities. However, keep in mind that more microscopes and/or volunteers will be needed.

Offering Some of the Learning Station Activities at a Time

— *Rotate classes.* A festival room is set up with a different subset of the stations on day one, two (and possibly on more days). Several classes a day visit the festival room.

— *Rotate stations.* Two (or three) teachers divide up the learning station activities, each setting up four to six stations in their classroom for two or three days. Later, the teachers swap stations. This could be repeated by successive pairs of teachers.

If your time is limited to 45–60 minute class periods, one of these two formats may be best for you. There are many different combinations of four to six learning stations that you can make. Depending on the goals and schedule, several classes could each spend a class session in which students visit four to six stations. This format requires fewer volunteers at a time and less space than if all 10–12 stations are set up at once. However, you will need to plan for volunteers and space over a period of several days.

Getting by with fewer microscopes (or having more microscopes per station):

Reducing the number of microscopes involves careful attention to which stations are set up at one time, and works best if you rotate classes to a festival room with the stations set up. Suppose, for example, you plan a two-day festival. You plan to schedule several classes of about 32 students to rotate through a room with six stations each day (about five or six students per station).

You might plan to have six stations on one day and the other six the next. The trick is to divide up the stations so the four stations that don't need microscopes (Up Close, Fingerprint Ridges, and the two second activities for Salts and Pond Life—if you are including them) are evenly divided between the days. That way, on each day, only four of the six stations require microscopes. With two microscopes per station, you could get by with eight microscopes; with three per station, 12 microscopes. With 16 microscopes, you could have four per station!

If you choose the second format above, and with stations rotating to different classes, and all the station activities are being used at the same time, the total number of microscopes is not reduced. However, with careful planning and a longer time frame, stations could be rotated to classes in sequence, so fewer microscopes would be needed.

Formats that use only some stations at a time can be adapted to meet the following priorities:

- **If your priority is introducing as many students as possible to the learning stations:** What if you can't hold a large group festival all at once, but want to introduce multiple classes to *Microscopic Explorations*? Say, for example, you have 45–60 minute class periods with five or six different classes a day. Students could rotate every ten minutes and visit four to six stations per class period. This way, it would take two or three days for all those classes to visit all the stations.

- **If your priority is providing more time at each station:** If you want timed rotations of students to stations, there are two main ways to extend the time students have at stations. One way is to schedule longer sessions of up to 90 minutes. If you can't lengthen the class session, have students visit only **two** of the four to six available stations set up in the room during each class session. This way, there are enough learning stations to accommodate the whole class, but students spend half a class session on each station (about 20–30 minutes). This option further extends the number of class sessions needed overall; it would take more days for everyone in the class to visit 10 stations.

- **If your priority is having multiple stations of certain activities:** By making multiple learning stations of some of the activities, you can have more of the class focusing on fewer explorations at a time. For example, you might decide to create a session in which you present only three of the 10 learning station activities. By producing two stations of each, you have six stations. There would be double the materials to gather (for instance double the insects or fabric samples), but the number of microscopes and volunteers wouldn't be more than other classroom options. Once again, it would take more sessions to complete all 10 learning stations this way.

- Creating six stations of a certain activity transforms the learning station format into a whole class activity for a specific purpose, curricular or practical. For example, suppose you want to present the GEMS *Fingerprinting* unit, and then have everyone do the Fingerprint Ridges station activity. For that day of your festival, all the stations could be Fingerprint Ridges. Or, suppose all your brine shrimp have just hatched, and you want to use them in a relatively short time! If you have enough microscopes, depression slides, and other materials, you could set up six Brine Shrimp stations and give students a longer time for this activity.

Offering a Single Learning Station at a Time

You may wish to use individual stations separately as an activity for a science learning center in your classroom, or as an introductory activity for a unit in your curriculum. If so, set up the station and rotate students through. You may also want to use a station as a place for students to visit when they have finished their work early.

These adaptations work whether you rotate stations to groups of students or students to stations in a room, and whether or not you must get by with fewer microscopes.

A teacher who used this option with multiple classes for five days in a row says, " We were all tired by the end of the week. Next time we will take a break for a day or so."

About Microscopes and Volunteers

For more information about microscopes, please see "Behind the Scenes," "Selecting School Microscopes," and the preparatory whole-class session beginning on page 27.

Microscopes

Generally, microscopes with low power are easier and more fun to use, and are best to view the wonders of the *Microscopic Explorations* Festival. A magnification of 40x or lower is best. If possible, we recommend that half of your microscopes be low-power microscopes called dissecting microscopes, and the other half be compound microscopes. With compound microscopes, this usually means using the least powerful (shortest) objective. Again, please see the preparatory whole-class session and "Behind the Scenes" for more information.

If you don't have dissecting microscopes, don't despair—you can hold a successful festival using only compound microscopes (which are much more likely to be available). If you have only a few dissecting microscopes, you might want to put them at the learning stations where students observe small organisms (the Small Creatures, Brine Shrimp, and Pond Life stations). A possible alternative to dissecting scopes is a set of inexpensive 30x handheld "flashlight" style magnifiers. Many schools have microscopes, sometimes tucked away in corners or cupboards, in various stages of repair. Even if a particular school does not have a classroom set of microscopes, someone in the district may know where and how to locate enough for the learning stations. A nearby high school may be a possible source for borrowing microscopes. Please see pages 115–119 for more information on microscopes and how to gather what you need.

Microscopic Explorations was developed with the support of and in close collaboration with the Microscopy Society of America (MSA), and from its beginnings was envisioned as a means of building partnerships between teachers and scientists who use microscopes in their work. **For that reason, a listing of MSA regional and local chapters is included on page 104. MSA members may be able to assist you in finding microscopes to borrow if your school does not have its own, and advising on small grant proposals to obtain your own.** They may also be able to repair or clean microscopes that you think are beyond hope. Based on their understanding of the need for high-quality science and math education, they can also serve as scientist-volunteers who help out with the festival or come to talk to the class about their work.

Volunteers

The Salts learning station's additional activity, Crystal Suncatchers, must have an adult present for safety reasons. All of the stations, however, benefit from having someone tend to necessary logistical details, encourage student questioning and investigation, and help keep students on task. Having a volunteer at each station makes a *Microscopic Explorations* Festival run smoothly and ensures that students maximize their learning. It also allows you as the teacher to concentrate on student learning rather than on station logistics. In a large group festival, 10–12 volunteers are needed for 90 minutes or more. In classroom festival formats where only some of the stations are available at a time, some teachers have managed to get by with only one or two volunteers at a time, for all or part of several days.

Who Can Help?

Some of the scientists you've contacted for microscopes may be enthusiastic volunteers, or could come as special guests after your students' curiosity has been stimulated by experiencing the festival stations. Parents are also great volunteers and technical expertise is not necessary to be a very effective volunteer. If most of the parents in your community work at the time you are planning to hold your festival, consider inviting grandparents or contacting a local branch of the American Association of Retired Persons. It has been our experience that most adult volunteers enjoy the festival as much as the students do!

Peer Teaching

In some cases, teachers have had their students present the festival to other students or parents after they have experienced the stations themselves. This can be a particularly effective means to do peer teaching and develop a variety of skills. This is also a way to form a bridge between middle school and elementary school students.

Volunteer Guidelines and Orientation

To help you get the best help from your volunteers, we have included Volunteer Sheets for each learning station activity. These briefly explain the learning goals of each station, suggest additional questions to ask students, and outline how to maintain the station so it is safe and ready for the next group of students. There is also a sign at each station with reminders to help students use microscopes safely and successfully.

For the most part, the adult help you need requires no special preparation, but you may find someone who interferes with students' open-ended investigations by telling or showing the students too much. Sometimes a scientist or technician who lacks understanding of guided-discovery learning may launch into lecture mode. One side of the Volunteer Sheets emphasizes the goal of helping students make their own discoveries.

A short orientation with your volunteers about how to guide student learning is invaluable. This may take place on the day of the festival or when you are recruiting them on the phone. You can reinforce this at the festival itself by telling the students (while your volunteers are listening), "We adults may help to remind you about the goal at each station, but you will be the scientists today. You are the ones who will investigate each problem and make your own discoveries." Ideally, you will want to have another person take on the responsibility for lining up volunteers. Perhaps a parent or interested community member would coordinate that task for you. If so, they should pass on the primary idea of letting students make their own discoveries to everyone they recruit.

The North Carolina Museum of Life and Science produced "Sharing science with children" as two excellent pamphlets in the early 1990s. They have now been reprinted, and can be ordered (the first five copies are free) from the North Carolina Museum of Life and Science, PO Box 15190, Durham, NC 27704.) Both are available in downloadable form on the web:

Sharing science with children—a Survival Guide for Teachers http://www.noao.edu/education/ncmlstg.html

Sharing science with children—a Survival Guide for Scientists and Engineers http://www. noao.edu/education/ncmlssg.html

What You Need (for the entire festival)

General Supplies

- ❏ 24 manila folders, letter size (for making signs, water drop magnifiers, and hole-punch slides)
- ❏ 1 container rubber cement
- ❏ 2 bottles white glue
- ❏ 2 large sheets butcher paper: 1 white, 1 blue
- ❏ several pieces of construction paper in a variety of light colors (for Discovery Quilt), plus black (for Activity 5)
- ❏ access to a paper cutter
- ❏ newspaper with small print
- ❏ book or page with a large printed word
- ❏ five packs or rolls paper towels
- ❏ 1 hole punch
- ❏ 2 packs crayons (or permanent felt-tipped markers)
- ❏ a variety of markers, pens, pencils (for the Discovery Quilt)
- ❏ 1 wide-tip marker
- ❏ 1 roll masking tape
- ❏ 4–6 sharpened No. 2 pencils
- ❏ 3 trash containers
- ❏ a few 3" x 5" index cards
- ❏ 1 sheet of 8 ½" x 11" paper
- ❏ scissors
- ❏ 1 yardstick

Consumables

- ❏ 1 copy of "The Parts of a Compound Microscope" student sheet per student
- ❏ 1 copy of "The Parts of a Dissecting Microscope" student sheet per student
- ❏ 1 copy of the student booklet per student
- ❏ 60–100 clear plastic deli-type container lids
- ❏ 1 roll duct tape (for taping down electrical cords)
- ❏ 1 roll waxed paper
- ❏ 12 rolls clear (**not** Magic™) transparent tape, with dispensers
- ❏ 5 rolls ¾"-wide Magic™ tape, with dispensers
- ❏ about a tablespoon of each:

___ granulated white sugar	___ cream of tartar
___ laundry detergent	___ alum
___ white flour	___ rock salt
___ cornstarch	___ boric acid
___ baking soda	

- ❏ about 5 tablespoons of table salt (for Activities 5, 7, and 9)
- ❏ 2 packages Epsom salt
- ❏ 30 pieces of white scratch paper
- ❏ 1 container brine shrimp eggs
- ❏ (optional) a sprig of Elodea aquarium plant
- ❏ 1 oz. live brine shrimp adults
- ❏ some newly hatched brine shrimp larvae
- ❏ pond water with living organisms
- ❏ several dozen pipe cleaners
- ❏ 50–60 styrofoam peanuts
- ❏ 20–30 popsicle sticks

- ❑ 20–30 straws
- ❑ 20–30 toothpicks
- ❑ 2 small squeeze bottles of liquid dishwashing detergent

Non-Consumables

- ❑ 1 two-sided copy of each of the Volunteer Sheets
- ❑ 1 copy of each of the station signs
- ❑ 8 copies of Microscopes: Tips for Safety and Success
- ❑ 24 small ziplock bags (qt. size)
- ❑ 40–60 hand lenses
- ❑ microscopes (compound and dissecting, as available)
- ❑ extension cords
- ❑ power strips
- ❑ table lamps (a short fluorescent tube is particularly good)
- ❑ several empty clear vials, pill bottles, or water bottles with lids, any size
- ❑ several clear, flat-sided plastic or glass containers, any size
- ❑ 2 large, clear glass or plastic containers with curved sides (such as two-liter soda bottles)
- ❑ (optional) 2 large, clear glass or plastic containers with straight, flat sides
- ❑ 8–10 clear, flexible plastic cups (about 6–10 oz. size)
- ❑ 8–10 medicine droppers
- ❑ 8–12 plastic depression slides
- ❑ about 10 clear, flattened marbles
- ❑ about 10 clear, round marbles
- ❑ 6–12 small assorted fabric samples (including woven, knitted, and pressed)
- ❑ (optional) several pieces of Velcro
- ❑ 2 empty deli or cottage cheese containers with lids
- ❑ a few empty yogurt containers or film canisters
- ❑ a box, tray, or paper plate (for storing insects)
- ❑ tweezers
- ❑ 1 one dollar bill
- ❑ 1 postage stamp
- ❑ 1 penny
- ❑ 1 film negative
- ❑ a few black and white photos
- ❑ a few new color photos
- ❑ 1 black and white newspaper picture
- ❑ 6 or more colored newspaper pictures
- ❑ a few business cards
- ❑ a few magazine pictures
- ❑ (optional) a few Georges Seurat pictures
- ❑ 3–6 labeled sand samples from several locations, one tablespoon each
- ❑ world map or globe
- ❑ small Post-it® Notes or adhesive dots to mark locations on map
- ❑ (optional) a few rock samples, a few shells, a few coral samples
- ❑ 1 hot pot
- ❑ 1 metal teaspoon
- ❑ 6–10 dead insects, spiders, and parts

Getting Ready

Whether you choose a large group or a classroom festival, we strongly recommend that the tasks of rounding up and preparing materials, and gathering volunteers and microscopes be shared by two or more adults!

One month before the festival:

1. *Make some general decisions on the format and schedule of your festival.* Having read "Planning Your Microscopic Explorations Festival," choose the festival format that will best meet your needs. Familiarize yourself with the ten learning station activities by reading the short introduction to each activity, the station signs, and the Volunteer Sheets, which include tips on the logistics of each station. Agree on a plan with other teachers whose classes will be enjoying the festival, and divide up some of the preparation tasks. Delegate as much as possible of the preparation to volunteers.

2. *Schedule the use of the room.* Once you have decided where and when you will hold the festival, you may need to sign out the room. Do this as far ahead of time as possible, and before you do any recruiting of volunteers. Be sure that the room will be available for at least several hours before and an hour after the time you plan to hold the festival. Check the locations of electrical outlets and whether they are for two- or three-pronged plugs. (Power strips can be helpful in making connections.)

3. *Identify and recruit volunteers.* Get help from other teachers, parents or aides for recruiting volunteers if you can. You will need at least 10–12 volunteers for the large festival, but schedule a few extra, to allow for last-minute cancellations. For a format with fewer stations, plan for about one volunteer per station for each day of the festival. Tell volunteers what the festival is about, the kind of help you will need, and how much time is required of them. Let them know when and where it will take place. Remember to put them at ease by telling them you do not expect them to be experts on microscopes, and let them know that all the instructions will be written out for them on the day of the festival.

4. *Locate (and reserve if necessary) microscopes, lights, power strips, and extension cords.* If you need to borrow any of these items, make the arrangements early enough so you will be sure to have enough on hand. Eight of the learning stations each need at least two microscopes. Be sure that you have enough desk or table lights for microscopes that don't have their own built-in lights, and enough extension cords and power strips for each light or microscope that has a built-in light source.

If any of the microscopes are battery powered, you may want to have some spare batteries on hand.

5. *Begin gathering and sorting the materials for all the stations.* This is a great task to delegate to a parent or other volunteer. We have included a "Sample Letter to Parents" (page 149) that includes a list of many—but not all—of the needed materials. Or, you could duplicate the complete list on pages 14 and 15 and attach a short note home, or ask an active and organized parent or group of parents to coordinate the collection process. As you or the volunteer collect the fabric scraps, sand, and other items, they'll need to be organized by learning station. Most of the station materials are common and easy to find.

6. Locate or order the following important, but less common items: *(Please see "Sources for Materials," page 120, for additional information.)*

a. Live brine shrimp. Check delivery schedules to see which days live brine shrimp are stocked at a local aquarium store—most only stock them a couple of days a week, and they die quickly. If you do not have a local source for about one ounce of live brine shrimp, you may need to order them from a scientific supply company (see "Sources for Materials").

b. Pond organisms. You can usually find pond water somewhere in your community. Look in ditches, lakes, or fish ponds. Standing or slow-moving water is best, though not green water, which will not have many animals or much diversity. Be sure to scrape the mucky areas on the bottom and sides of the pond; this is where many of the most interesting tiny organisms live. Check a sample with a microscope to be sure there are living organisms moving around. (You may see, for example, small creatures, either darting to-and-fro or moving slowly and steadily; they could be light or dark-colored, rounded or bug-like.)

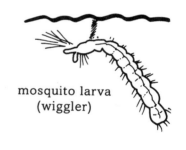

Tubifex worms

Setting out a bucket of plain water with some hay or grass in it for several weeks to a month, especially during spring and summer, will also provide an interesting sample. If you live in an area that freezes over during the winter, you may want to start an indoor aquarium of pond water so you will have a good supply whenever you want to hold your festival. (If you already have a fish tank, pump, and aerator, they would be very helpful to keep the water oxygenated so organisms thrive, and to prevent blooms of algae that will harm the organisms.)

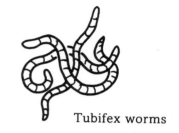

mosquito larva
(wiggler)

If you do not have a good source of pond water organisms, you may need to order pond organisms such as *Daphnia* from a scientific supply company. Local aquarium stores may carry *Tubifex* worms, and mosquito larvae may be available from a mosquito abatement agency. If your pond does not have many plants, you may want to buy *Elodea* or other aquarium plants from an aquarium store.

Elodea plant

If you live near the coast and can collect marine organisms, you may add these to the Pond Life station. A good way to do this is to locate a marina or dock and scrape off several of the organisms and plants growing on the bottom and sides. (These samples will need to be collected the day of or the day before the festival so they will stay alive.)

Daphnia

c. Dead insects, spiders, arthropods. You'll need six to ten dead "bugs" for the Small Creatures learning station, and it's good to have a few extra in case of loss or damage. Ask for student volunteers to collect dead bugs. It's fine if some students are not eager to volunteer for bug collecting; you'll probably have other students who are very enthusiastic. Tell them not to kill bugs! Explain that they are to collect only bugs that are already dead, like the ones often seen along windowsills. It would be great to include a fly, a moth, a fruit fly, and a spider in the collection. Other creatures of interest would be ants, beetles, butterflies, roly-polys (isopods), crickets, or grasshoppers. Molts are also interesting. Also, a spider web is a good source of interesting bugs/insects and/or their remains.

Mention that flies and some other insects can carry germs. **Be sure that students know that they must not touch the specimens with their hands!** Give each student volunteer an index card, a couple of toothpicks, and a plastic container with a lid to take home. (Film canisters or yogurt containers work well.)

Have them fold the index card in half widthwise once. Demonstrate how to use the corner or folded area of the index card, along with the toothpicks, to scoop up any specimens they find, carefully slide them into the plastic container, and put the lid on. **Tell students to wash their hands after collecting.**

You might mention to the collectors that smaller specimens can be the most interesting to magnify. For instance, a delicate fly or mosquito may be easier to observe with a microscope than a big, solid beetle. Individual legs, antennae, or wings are also excellent to observe. To supplement your collection, check for dead crickets at local pet stores that sell them. Fruit flies should be easy to find at a produce store.

d. Depression slides for the Brine Shrimp and Pond Life stations. Depression (or "well") slides have a central "dip" for water or other liquids. Depression slides are great for observing tiny living creatures or other live materials suspended in liquid. You'll need about six plastic depression slides for each of these two learning stations. For safety reasons, and because for these stations plastic slides work well, we don't recommend using glass slides. Plastic, however, does scratch more easily than glass, so you may need to get a few extra to replace scratched ones.

e. Clear plastic lids for the suncatcher activity. You'll need one lid for each student who visits the Salts station's second activity, Crystal Suncatchers. You could collect clear, plastic deli-lids, buy some at large stores such as "Smart and Final," or order them from a restaurant supply company. (See the "Sources for Materials" section.) The diameter of the lids can vary, from the size found on individual yogurt containers to the larger deli-container size. The lids do have to be clear, not have any holes in them, and need to have a shallow depression into which you can pour a teaspoon or so of salt solution.

You may be able to order plastic lids through your school cafeteria.

f. Magnifying (hand) lenses for all the stations. You'll need about 50 hand lenses for the ten stations. If you can't gather that many at once, you might need to order some.

g. About eight medicine droppers. Plan to borrow, gather, or order ahead.

Two weeks before the festival:

Send out reminder letters to volunteers. The reminder letter can include information about how to get to the school and room, with a simple map, if needed. If you know which station each volunteer will be responsible for, you may want to send out a copy of the two-sided Volunteer Sheet for them to review ahead of time. Otherwise, they can read them on the day of the program. Remember to emphasize students making their own discoveries, and, if planned, to schedule an orientation session with the volunteers just before the festival.

One week before the festival:

1. Make station signs. Make one copy of the station sign for each of the learning station activities. You may want to copy the signs on colored paper. **Be sure to include the "Microscopes: Tips for Safety and Success" page on each sign where microscopes are used (see page 151).** Rather than being set flat on station tables where they can get lost or overlooked, the signs will need to stand up, be mounted on a wall, or hung. We recommend gluing them onto manila file folders, and laminating them for durability, as in the following directions:

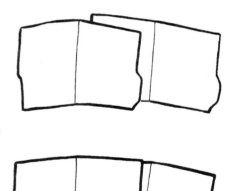

a. Glue two folders together so you have a three-panel sign. (Use any paper glue.)

b. Use rubber cement (best because the pages won't "bubble up" off the file folders) to attach copied signs to backing.

c. Laminate signs or cover them with contact paper (optional).

d. Label a tab or edge of the sign for easy retrieval from your *Microscopic Explorations* file.

2. *Make student booklets or provide journals.* For recording their observations at all 10 stations, each student will need three double-sided sheets (masters begin on page 152). By stapling these sheets together in the order shown below, you can create a booklet with a half-page of space for drawing and writing responses to the questions on each station sign. If you would like students to have additional space for recording observations, you might add an extra page or two of blank paper to each booklet before stapling. (Of course, assembling the booklets is something that a volunteer or students themselves can do. This might be an excellent task for a helpful group of parents.)

Instead of using the booklet pages in the guide, you can have students record observations and answer questions in journals. However, please note that each student will need a copy of the booklet page for Fingerprint Ridges (page 154). Make enough copies of that page for each student to have one, and have them available at the learning station.

Please note: When duplicating the page with the fingerprints, please test the setting on the copy machine to make sure the fingerprint patterns show clearly. A lighter setting may be best.

To make the student booklets:

a. Copy the six pages after page 151 back-to-back on three sheets of 8 ½" x 11" paper, ordered just as in the guide. In other words, the front and back cover page should be copied on the opposite side of the Up Close/Pond Life page; the Brine Shrimp/Fingerprint Ridges page should be copied on the opposite side of the Dots and Dollars/Small Creatures page; and the Kitchen Powders/Fabrics page should be copied on the opposite side of the Salts/Sand page. It may take a few tries at the copy machine to make sure all pages are oriented correctly—that the words are all right-side up when the page is flipped over.

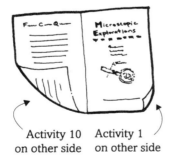   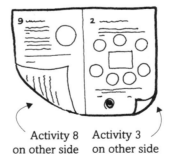   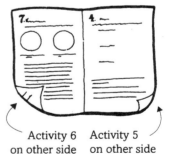

| Activity 10 | Activity 1 | Activity 8 | Activity 3 | Activity 6 | Activity 5 |
| on other side | on other side | on other side | on other side | on other side | on other side |

b. For each booklet, stack the three pages as shown in the illustration.

c. After double checking that the pages are in the correct order, fold the booklet in half.

d. Secure the pages together with two staples on the "spine" of the booklet. To allow the stapler to reach the spine, fold one side of the booklet temporarily. Although stapling in this way is a little more work, it allows the pages to open fully and lie flat.

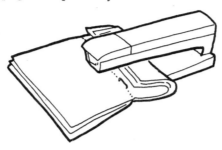

Note: In case you are not planning to feature all of the stations, you could re-design the booklet to include only those stations you are using, or you could simply cover up the stations you are not using with a blank sheet of paper, leaving more room for drawings or other student work.

3. Make hole-punch slides. For holding materials to observe with a microscope, scientists generally use flat, clear glass slides, usually with another thin glass covering called a cover slip. (This cover slip helps to hold things flat and stationary.) For the festival, we suggest you use instead a new kind of "home-made" slide called a "hole-punch slide."

Five of the learning stations (Sand, Salts, Kitchen Powders, Small Creatures, and the brine shrimp eggs at the Brine Shrimp station) use a total of about 56 hole-punch slides. We suggest you make labeled, prepared sets of hole-punch slides for all these stations ahead of time, plus some extra pre-punched pieces of manila folder to use in making additional slides, if needed.

To make the hole-punch slides:

a. Cut about 60–70 pieces of manila folder, 1" x 2".

b. Use a hole punch to make a hole in the center of each piece.

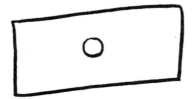

c. Cover the hole on one side with clear transparent tape.

d. **Turn over the slide so the sticky side of the tape in the hole faces up. Label the slide on this side. The sand slide labels should tell where the sand is from.**

e. Dip the slide into the appropriate powdery or grainy material (see each of the five station "What You Need" sections for materials) and brush it off gently. One way to do this is to place the powder or grains in a plastic ziplock sandwich bag, drop the slide in, close the bag, and shake. Some of the particles will remain on the tape. *Do not cover the hole with more tape;* the uncovered particles will be easier to observe, and will stay on the tape indefinitely. **Make two slides of each sample.**

f. The rock salt crystals for the Salts station may be too large for the dipping method. Instead, select one crystal of rock salt, and stick it on a slide, or use rubber cement to hold crystals to the slide. Keep some extra rock salt handy during the festival in case the crystals fall off the slides.

g. For the Small Creatures learning station, you may want to punch a cluster of two or three holes in the slide, both to provide more "sticking" space, and to allow more light through the slide. Enlarge the hole by punching extra holes lengthwise along the center of the slide, rather than clumping the holes together in the middle. (Because the tape tends to buckle with larger holes, a longer, thinner opening is better than a larger round one.) Use tweezers to place the small creature on the sticky tape. (See "Getting Ready" for the Small Creatures station.)

h. Place all slides for a particular station in a sandwich bag.

Two or three days before the festival:

1. Reconfirm volunteers. A telephone call is a good idea, since some may have misplaced their letters or lost track of the date or time.

2. Start brine shrimp eggs. Use a plastic container that holds about a pint. Label it "Brine shrimp larvae." Put about 3 tablespoons of salt (kosher is best, but table salt will work as well) into the container and fill it about halfway full with tap water. Stir until salt is completely dissolved, then add about ⅛ teaspoon of the brine shrimp eggs. Stir and shake vigorously. Keep in a warm, bright place. They should hatch and be ready for class within two or three days. Shaking the container periodically to aerate it may also help. *Please note:* **In some cases, even with careful effort, the brine shrimp eggs just don't hatch—some batches are more viable than others. If this happens, don't worry; the adult brine shrimp will be enough to make the station a success.** If you really want to get the eggs to hatch, try varying conditions such as temperature, salinity, and type of salt or water used, or you may need to buy new eggs.

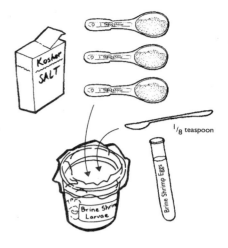

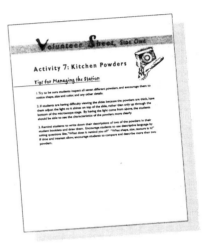

3. *Buy live brine shrimp adults from an aquarium store.* Buy 1 ounce. They will come in a plastic bag. Open the bag and set it into a hard plastic container such as a yogurt or ice cream pint or quart container, and keep the brine shrimp in the refrigerator until you are ready to use them. They will last a few days; probably not more than about a week. Be sure to open the bag to give them oxygen!

4. *Make Discovery Quilt chart.* Label a piece of butcher or other chart paper, "Discovery Quilt," and draw a grid of squares approximately 4" x 4" (or whatever dimensions are best for you) for students to use for their individual squares. Cut an assortment of colored paper into the same size squares/shapes—again the exact size and shape are not critical.

5. *Photocopy the Volunteer Sheets.* You'll need one two-sided sheet per station. Side one of the sheet is different for each station. Side two (page 150) has general tips, and is identical for all stations. After copying all the ten different station Volunteer Sheets, copy the "side two" page on the backs of all ten.

6. *Gather microscopes, lights, power strips, extension cords, tape for taping down cords, world map or globe, cafeteria trays, hot pot, and other equipment.* Be sure these items are put in a safe place, especially if you are borrowing them. Make sure that everything works! If you can set up the room a few days ahead of time, go ahead, but you probably will not have that luxury!

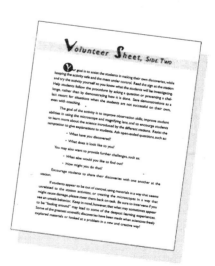

One day before the festival:

Collect organisms from pond and/or ocean. From the areas you checked weeks earlier, collect about a gallon of pond water in a plastic container. Be sure to scoop some muck from the bottoms and sides of the pond to add to the water. Look to be sure that there are some animals flitting around in the water. (See page 17, under "One month before the festival," 6b, for reminders on getting the best samples.)

If you want to collect marine organisms, the easiest way is to go to a nearby marina and scrape off seaweed and attached animals from the bottom of the docks. Take a metal strainer and use it to scrape off a chunk, then scoop it up into a plastic bucket or other large container. Keep pond and ocean samples in the refrigerator overnight, with no lid on the container.

On the day of the festival:

1. Set up the stations around the room. Setting up all the stations will take about an hour, or perhaps more the first time you do it. Ideally, each station will have space equal to a large cafeteria table, or about six feet by three feet. For smaller groups, clusters of desks, smaller tables and countertops can make good learning stations. Of course, table stability is an important consideration, not only for the preservation of microscopes, but for the ease of viewing through microscopes without too much "wobble."

The Salts and the Pond Life stations should each have two tables, if possible. Try to put the Pond Life, Sand, and Discovery Quilt stations near a wall where the murals or maps can be posted. Stations that require a microscope may also need access to an electrical outlet, depending on the type of microscopes or lights you are using. The Salts station will require an electrical outlet and an extension cord for the hot pot.

The best way to arrange tables so cords are out of the way is with the narrow end against a wall near an electrical outlet. Plug in any necessary light sources or microscopes and be sure to tape all extension cords down securely. Put short pieces of duct tape across the cord at intervals, rather than taping the whole cord down. Avoid loose coils.

Put materials out as described in the instructions for each station. Be sure to include the station signs. For all stations with microscopes, adjust the microscopes and lights so they are focused and ready to use.

2. Orient your volunteers. Have your volunteers come at least half an hour early so you can go over the stations with them. Let them familiarize themselves with the materials, station signs, student sheets, "Microscopes: Tips for Safety and Success," and both sides of the appropriate Volunteer Sheets. Ask if there are any questions. Give them an overview, including your objectives for the festival, some advice for how to guide students yet allow them to make their own discoveries, and information about logistics such as timing, number and grade level of students, and clean-up procedures.

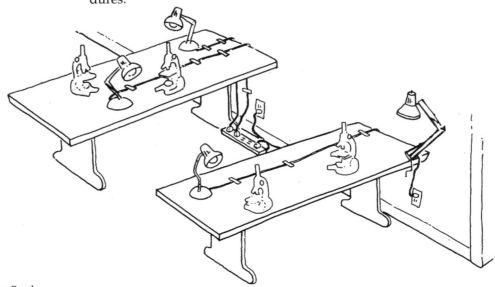

Preparation Checklist

One month before the festival:

___ 1. Make decisions on format and schedule. Agree on a plan with other teachers and divide up preparation tasks. Delegate as much as possible to volunteers.

___ 2. Schedule the use of the room. Check locations and types of electrical outlets.

___ 3. Identify and recruit volunteers.

___ 4. Locate and reserve microscopes, lights, power strips, extension cords.

___ 5. Begin gathering and sorting the materials for all the stations.

___ 6. Locate or order:
- ❐ live brine shrimp
- ❐ pond organisms
- ❐ dead insects, spiders, arthropods
- ❐ depression (or "well") slides
- ❐ clear plastic lids for suncatcher activity
- ❐ magnifying lenses
- ❐ eight medicine droppers

Two weeks before the festival:

___ Send out reminder letters to volunteers.

One week before the festival:

___ 1. Make station signs. Include "Microscopes: Tips…" on each sign.

___ 2. Make student booklets or provide journals.

___ 3. Make hole-punch slides.

Two or three days before the festival:

___ 1. Reconfirm volunteers.

___ 2. Start brine shrimp eggs.

___ 3. Buy live brine shrimp adults from aquarium store.

___ 4. Make Discovery Quilt chart.

___ 5. Photocopy Volunteer Sheets, both sides.

___ 6. Gather microscopes, lights, power strips, extension cords, tape, world map or globe, trays, hot pot, and other equipment.

One day before the festival:

___ Collect organisms from pond and/or ocean.

On the day of the festival:

___ 1. Set up the stations. Two tables for Salts and Pond Life stations, if possible.

___ 2. Orient volunteers. Have them come at least half an hour early.

HAVE A GREAT LEARNING EXPERIENCE !

"...by the help of Microscopes, there is nothing so small, as to escape our inquiry; hence there is a new visable World discovered to the understanding."

Robert Hooke, **Micrographia**, 1665

Preparatory Whole-Class Session(s)

Magnifiers and Microscopes

Overview

This session is made up of two parts. Part 1 is a guided discovery activity about the properties of lenses. Part 2 is a short introduction to microscopes, with time to label a diagram of a microscope. Depending on your schedule and the age of your students, you may want to break these two parts into two full class sessions, or give the microscope labeling activity as homework.

Magnifying lenses work by "fooling" our eyes and brain. The angle of the light rays traveling toward our eyes is bent in a way that makes us perceive the object as taking up a larger space than it really does. The more steeply curved a magnifier is, the more it bends the light, and the more it will magnify. You'll want to read the "Special Section on Optics" for teachers (pages 123–129) to learn about how lenses magnify, but **resist the temptation to provide this information to students before the water drop activity.** *After* students have discovered for themselves the magnifying property of water drops, they will be much more receptive to learning more.

Please note: Depending on the experience of your students, it is possible to skip one or both parts of this whole-class session. Up Close, one of the ten learning station activities, will "re-focus" on the concepts introduced in the first part of this session. At that station, students explore the magnifying properties of various objects and create portable water drop magnifiers. Some teachers let the Up Close station serve as an introduction to lenses and magnification, and opt to skip this preliminary whole-class session on magnifying lenses—presenting only Part 2, on microscopes. Teachers whose students already have had recent experience with microscopes might opt to go straight to the festival, or, if they think it would be helpful, present only Part 1 on magnifiers.

For more information about compound and dissecting microscopes, please see "Compound and Dissecting Microscopes" on page 114 and "Selecting School Microscopes" on page 115.

Glass half marble on a leaf.

Part 1: Magnifiers

What You Need *(for Part 1)*

For the class:
- ❑ 1 magnifying lens
- ❑ a book or page with a large printed word
- ❑ 2 large, clear glass or plastic containers with curved sides (such as two-liter soda bottles)
- ❑ *(optional)* 2 large clear glass or plastic containers with straight, flat sides

  Note: Not many products come in large, clear containers with flat sides, so these may be difficult to find. (Some bottled water comes in square containers.) While having two flat-sided containers provides a nice contrast to the curved containers in "Introducing Magnifiers" (#5 below), they are not absolutely necessary.

For each group of four to six students:
- ❑ 1 plastic cup or other small container

For each student:
- ❑ 1 4" square of waxed paper
- ❑ 1 4" square of newspaper with small print and no pictures

Getting Ready *(for Part 1)*

1. Half fill a cup with water for each group of students. Cut up enough squares of waxed paper and newspaper for each student to have one of each. Have the student materials handy for quick distribution after your introduction.

2. Fill one of the large, curved containers with water and leave the other empty. (Make sure to remove labels from bottles.) Have these containers where everyone can see them during your demonstration.

3. If you have also obtained two large **flat-sided** containers, fill one of these with water and leave the other empty. Place these near the large curved containers.

Introducing Magnifiers

1. Demonstrate how to set a piece of waxed paper on top of a piece of newspaper, dip your finger into the cup, and shake *just a few* drops off onto the waxed paper. Ask students to take a few minutes to discover all they can about water drops.

2. After students have had a few minutes to explore, collect the materials.

3. Regain the attention of the whole class, and ask what they discovered about water drops. Accept all observations. [water forms little domes,

drops join together, drops can be broken apart, can be dragged] If no one mentions it, ask how the printed letters look through a drop. [The print seems bigger through the water drop.] Some students may notice that the smaller, more steeply curved drops have greater magnification.

4. Ask what shape the water drop is. [dome-shaped] Ask what they think caused the letters to look bigger. Explain that the curved water drop **bends the light so the print seems to be larger.**

MICROSCOPIC EXPLORATIONS

5. Draw the students' attention to the two (or four) large, clear containers. Point out that one (or two) of the containers is filled with water. Ask for predictions about which container will magnify printed letters. Hold a book or page with a large, printed word behind each container for the class to see. Ask for their observations. [Only the rounded container with water magnifies.]

> *If some students seem to have difficulty concluding that it is the **curve**, not the water itself, that is responsible for the magnification, you could ask them to compare the view through the curved side of the container with the view from the top. Only the curved side magnifies, not the flat surface on top of the water. For a class activity, you could put water in a clear plastic cup, one per group of four students, and have students compare the view through the water from the side (curved) and from the top (flat).*

6. Explain that *lenses* bend the light in the same way as the curved water drops and the curved container of water did. Ask the students to think of familiar things that use lenses, and list them on the board. [eyeglasses, hand magnifying lenses, cameras, microscopes, telescopes, projectors]

7. Circle hand lenses and microscopes on the list. Tell the class they'll soon be participating in a *Microscopic Explorations* Festival, and they'll get to go from one table to another, discovering interesting things with the help of hand lenses and microscopes.

8. Hold up a hand lens and mention that it is curved, like the water drop. Explain that microscopes have several lenses in them, working together to magnify powerfully. Hand lenses and microscopes both make objects appear bigger to our eyes by bending the light coming from an object.

> *If your students have had little experience with hand lenses, you might want to add a class session before the festival to let them use a hand lens to examine a page of print (and their clothes, skin, desktops, and whatever else is handy). Encourage free exploration and help students learn to focus by moving the lens back and forth between the object and their eyes.*

Part 2: Microscopes

What You Need *(for Part 2)*

For the class:
- ☐ 1 microscope (or a sample of each type they'll be using)
- ☐ a chalkboard or overhead projector
- ☐ a sample hole-punch slide
- ☐ *(optional)* 1 overhead transparency of "The Parts of a Compound Microscope" and "The Parts of a Dissecting Microscope" student sheets (masters on pages 32 and 33)
- ☐ *(optional)* 1 overhead transparency of a drawing of any other type of microscope you might be using

For each student:
- ☐ 1 "The Parts of a Compound Microscope" student sheet (master on page 32)
- ☐ 1 "The Parts of a Dissecting Microscope" student sheet (master on page 33)

Getting Ready *(for Part 2)*

1. Make one copy of "The Parts of a Microscope" student sheets (both compound and dissecting microscopes) for each student. (You may also want to make an overhead transparency of each sheet to use during your introduction.)

Although the illustration on the student sheet shows one, many dissecting microscopes don't have a mirror.

2. Sketch the same type of a microscope on the chalkboard, without labeling the parts.

3. Have a microscope handy to refer to during your introduction.

Introducing Microscopes

Refrain from presenting an exhaustive lesson here on all the parts of the microscope, since students will have plenty of time to learn more during the festival. The goal for now is just to introduce the basics!

1. Point out the real microscope you've brought in, and ask who has used microscopes before.

2. Say that microscopes are great tools for seeing small things better, and in the coming days, everyone will get to enjoy using them. Explain that first there are a few basic things everyone needs to know about microscopes.

3. Draw your students' attention to the diagram of the compound microscope on the chalkboard or overhead, and tell them that this is called a *compound* microscope, meaning one with two lens systems—eyepiece and objective. Briefly point out the *eyepiece* lens and the *objectives*, mentioning

that there are several lenses within each tube called an objective. (You may want to mention the parallel grammatical terms *compound* words or *compound* sentences.)

4. As you introduce the various parts on the microscope(s), you might want to mention any major differences in the ones students will be using. For instance, the microscopes you use might have two objectives, rather than three.

5. Hold up a hole-punch slide, perhaps with a fly's wing on it, and explain how the slide was made. Explain that the hole lets the light through, and the sticky tape keeps the wing (or whatever you're looking at) in place. Ask where the slide should go on the microscope. [the *stage*]

6. Point out the hole in the stage, which lets light through, and the *mirror* or light below it. Right below the stage is a disk with large and small holes called a *light control.* Tell students the disk can be turned to let more or less light come up through the stage. (Usually, tilting the mirror will have the biggest effect on the amount of light.)

7. To get the clearest view, they'll use the *focus knob(s)* on the side of the microscope. Point out the knobs.

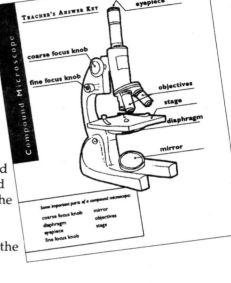

8. Now, point out the two (or more) *objectives.* The longer the objective, the more it magnifies. Tell students the shortest objective will usually give the best view of what they will be looking at. (With younger students, you may want to have them use only the shortest objective.)

9. If your students will be using dissecting microscopes, you may want to project the student sheet of a dissecting microscope on the overhead, and point out the lenses and stage. Explain that this type of microscope usually has low magnification (which is often very useful!). You may also want to mention that compound microscopes turn the image upside down and backwards, while in dissecting microscopes the image is right side up.

10. Distribute student sheets, and have students label key parts of compound microscopes, and, if you are using them, dissecting microscopes.

Going Further

1. Present the GEMS unit *More Than Magnifiers*, in which students investigate other devices that use magnifying lenses, including cameras, telescopes, and projectors.

2. Have students research the human eye and compare it to animal eyes.

Name: _____

Compound Microscope

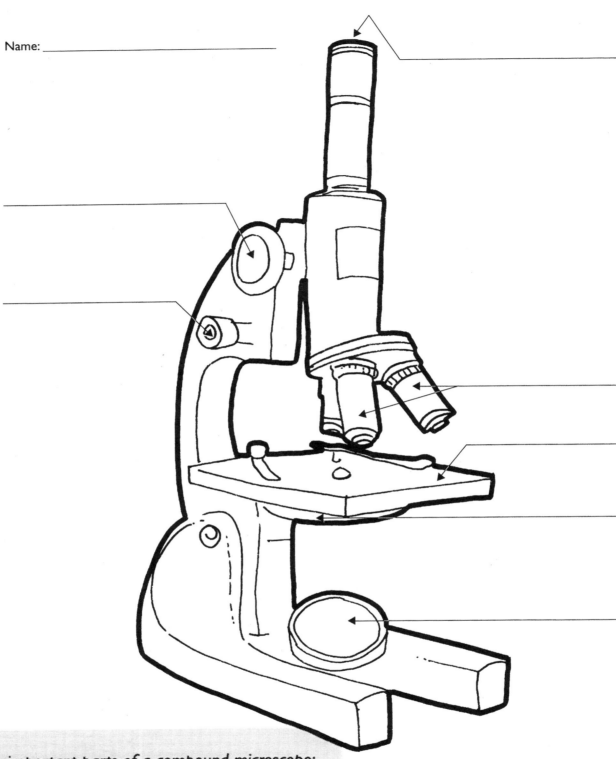

Some important parts of a compound microscope:

coarse focus knob   mirror

diaphragm   objectives

eyepiece   stage

fine focus knob

Name: _____

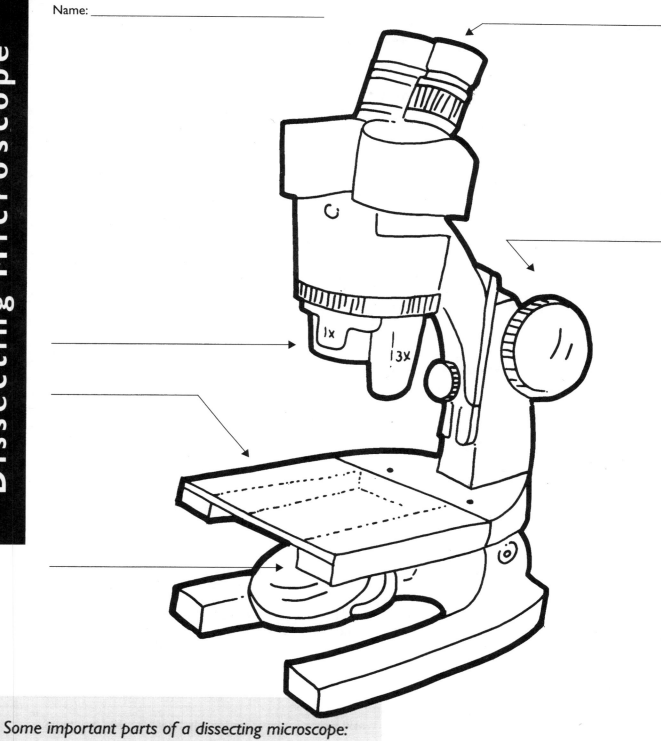

Some important parts of a dissecting microscope:

eyepiece   objectives

focus knob   stage

mirror

Learning Station Activities

Introducing the Learning Station Activities
(be brief!)

1. Tell students they are going to be making discoveries at different learning stations. Adult volunteers will be there to assist them but not to tell them an answer. Emphasize to students they will be the scientists today, the ones who investigate, make discoveries, and share their discoveries with others.

2. Say that their first task at each station is to observe something with hand lenses and microscopes. They should notice what the item looks like, observe it very carefully, and think of what it reminds them of, or other ways to describe it in detail.

3. Go around the room and **very briefly** introduce what they will do at each station. Explain that each station has a sign telling what to do and asking a few questions. Often the sign will ask them to draw and describe what they **actually see**—not what something is "supposed" to look like, but what they actually observe. Emphasize that great art isn't necessary.

4. Say that you'll assign them to their first station. Depending on whether or not you plan on timed rotations, give guidance on visiting the other stations:

No timed rotations:

- They will be working with a partner, and partners can visit the stations in any order.

- Each station has a limit of six (or whatever number you decide on) students at a time. If the station they are interested in already has six people, they should come back to it later.

- In addition to their booklets, they will have the opportunity to share their findings at the Discovery Quilt station. Say that at any time during the festival, they may go to that station and add their square to it, with either a discovery or question.

Timed rotations:

- They will be working in groups of six (or whatever number you have decided on).

- Explain the order in which groups will visit the stations and how much time they'll have at each station.

- Tell students they should stay at a station until your signal that it's time to change. Before each signal to change, you'll give them a two-minute warning so they can finish observations and/or recording, and tidy up the station for the next group.

5. Pass out the student booklets and explain that station sign numbers go with the ones in their booklets.

6. Remind students to clean up each station for the next group. Assign students to stations and have them begin.

Through A Lens Lightly: Ten Odes to the Microscope   by L.B.

I.

Ode to the mighty microscope
Allowing us at last to cope
With things that were before so small
We scarcely knew of them at all
Now we can look into the heart
Of matter and have made an art
Of microscopic inquiry
Who knows what wonders we might see?

II.

Through a lens lightly
Scope summons to see
Eagerly beckons
Curious me
Peering deep into
The spaces between
Network of structure
'Til now unseen
The objective is knowledge
Observation the key
Through a lens lightly
Scope beckons to me.

III.

The story of science
Is full of surprise
Results not the ones
We hypothesize
So many discoveries
Came to pass when
More questions arose
Then popped up again!
New understandings
Weren't found in a book
They happened because of
A much closer look.

IV.

Objective is knowledge
Observation the key
Open new windows
On reality
Somewhere a technician

Bends to her mission
Grinding a lens
Extending our vision.

V.

Creation of microscope
Telescope, magnifier
Helped science blossom
Great minds did inspire
The triumph of reason
The bending of light
The world of the atom
More stars in the night.

VI.

Today we can gather
Peer deeply inside
Find out about nature
World on a slide
Go to each station
In this micro-fest
Play our own roles
In the great human quest.

VII.

Look beneath surface
Enlarge grain of sand
Find lines and ridges
On human hand.
See shrimp and larva
Make observations
All can take part in
Great Explorations.

VIII.

Its power is awesome
Its image intense
The range of its vision
Surely immense
But for all of its magic
There's no hocus-pocus
It's simply a matter
Of bringing in focus!

IX.

On vast tree of science
Branches spread wide
Microscopic exploring
Wonder wide-eyed
Sights so amazing
Marvels untold
Tiny pulses of life
A joy to behold.
Physics, biology,
Chemistry, math
All mix together
Along the path
Crime lab to pond life
Fabric and flower
Metal and powders
Raised to high power.
All a broad vista
Horizon of light
Are tracks of life found
In a meteorite?
Objective is knowledge
Observation the key
Through a lens lightly
Scope beckons to me.

X.

These odes are a tribute
To great tube of light
That optically gifts us
Such fantastic sight
Whose lenses curve
Thus magnify
Multiply power
Of human eye.

As toward knowledge
People grope
Appreciate
The microscope!

"You can observe a lot by watching."

Lawrence (Yogi) Berra

Activity 1: Up Close

Overview

At this station, students investigate the properties of magnifiers. They construct their own water drop magnifying lenses, and also look through a variety of flat and curved objects to discover what magnifies an image. No microscopes are used at this station.

Note: This learning station activity reviews the properties of lenses introduced in the first part of the Preparatory Whole-Class Session. While most teachers present both, some opt to omit this learning station, and others keep this station and skip the portion of the preliminary whole-class session on magnifying lenses.

What You Need

❐ 1" x 2" pieces of manila folder—enough pieces for one per student plus a few extra
  These pieces of manila folder are similar to those made into hole-punch slides for some other station activities, but, at this station, students make them into little, hand-held water drop magnifying lenses. You can replace the manila folder pieces with pieces of clear, plastic covers from folders (often used for reports). This substitution eliminates the need for a hole punch or tape.

❐ 1 hole punch

❐ 1 roll transparent tape in dispenser (**not** the opaque or Magic™ kind)

❐ 4–6 hand lenses

❐ several transparent items, including ones with flat and curved surfaces, such as: spherical clear glass marbles, flattened clear glass marbles, small vials with tightly-closing lids, a clear, plastic water bottle (if flat-sided, clear vials or other containers, such as clear plastic specimen boxes—square or rectangular—are available, use these as well)
  Aquarium stores, florists, specialty gardening stores, and other suppliers often carry flat and curved clear glass marbles. Plastic vials and other clear, watertight containers might be donated by parents. See "Sources for Materials" for information on specimen boxes. The more such items you can gather the better, but 6 to 8 is enough.

❐ 4 plastic cups

❐ small newspaper pieces to look at—classified ads are best since the type is small

❐ 2 medicine droppers

❐ paper towels

❐ Up Close station sign

❐ Volunteer Sheet

Getting Ready

1. You will need two areas for this station. The first is for making water drop magnifiers. (To make one, students will punch a hole in a piece of manila folder, cover it with clear tape, and place a drop of water over the hole on the non-sticky side of the tape.)

2. In one area of the station, place tape, hole punch, and some of the newspaper pieces. Fill one plastic cup halfway with water, and put two medicine droppers in it. Put the manila folder pieces in another cup and set it nearby. (If you are using pieces of clear plastic instead of manila folders, you won't need to supply the hole punch or tape. Students can just place a drop of water on the clear plastic.) Have paper towels available in this area for wiping up spills.

3. In the second area, place the other newspaper pieces, and two cups filled with flat and round clear marbles. Fill some of the clear containers with water and leave some empty. Having a variety of flat and curved containers, both full and empty, helps students discover the properties of good magnifiers. To avoid having air bubbles in the filled vials, overfill them a bit using a medicine dropper before putting on the lids. You might even want to tape the lids on to avoid spills.

4. Place hand lenses in both areas for comparison. The station sign should be placed between the two areas of the station.

Going Further

1. Use the GEMS guide *More Than Magnifiers*, in which students investigate devices that use lenses, such as cameras, telescopes, and projectors.

2. Have students research the lens in the human eye.

Up Close

What to Do

1. Make your own hand lens using a drop of water.

Take a small piece of cardboard.

Punch a hole in it.

Put a piece of tape over the hole.

Put a drop of water on the non-sticky side of the tape.

Look through the water drop at some printing and see what you can see.

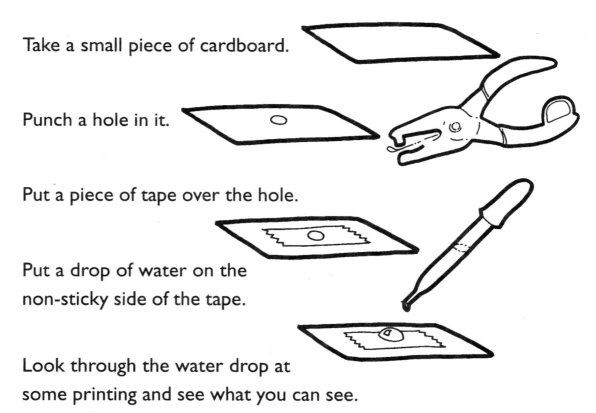

2. Try looking through all the other clear objects.

Activity 1
Up Close

Questions

A. Which objects make things look bigger?

B. How are the objects that magnify similar?

C. What makes a good magnifier?

Volunteer Sheet, SIDE ONE

Activity 1: Up Close

Tips for Managing the Station

1. Students may have trouble getting the tape over the hole in the piece of manila folder with no buckling or smudges. They may need a new piece to try again.

2. Be careful that students do not tip over the cup of water. If this happens, replace it and clean up the water.

3. If students need help with the medicine droppers, suggest the following:

   • Squeeze all water/air out of the dropper.

   • Lower the dropper into the water and "unsqueeze."

   • Lift the dropper, hold it over the slide, and gently squeeze.

4. The water drop magnifiers magnify only one letter or other small area at a time. Encourage students to move their water drop magnifier closer and farther from the newsprint to bring a letter into focus. Students may at first expect to see more, but they are usually quite happy with their magnifiers!

5. Encourage students to explore all of the objects and think carefully about what an object must be like for it to magnify.

6. Be sure students do not open the vials. If there are air bubbles in the vial, refill it to overfilling, using a dropper, and put on the lid.

© 1998 by The Regents of the University of California, LHS-GEMS. *Microscopic Explorations.* **May be duplicated for classroom use.**

"Beware of determining and declaring your opinion suddenly on any object; for imagination often gets the start of judgment, and makes people believe they see things, which better observations will convince them could not possibly be seen; therefore assert nothing till after repeated experiments and examinations in all lights and in all positions.

When you employ the microscope, shake off all prejudice, nor harbor any favorite opinions; for, if you do, 'tis not unlikely fancy will betray you into error, and make you see what you wish to see.

Remember that truth alone is the matter that you are in search after; and if you have been mistaken, let not vanity seduce you to persist in your mistake."

Henry Baker,
The Microscope Made Easy, 1742

Activity 2: Fingerprint Ridges

Overview

Of course, no two fingerprints are exactly alike. But what if a suspect's fingerprint is *very similar* to one found at a crime scene, and the suspect claims not to have been there? That's when crime lab scientists have to focus on tiny details of the two fingerprints to be confident that they really do match exactly. Experts use magnifying lenses to study individual lines, or "ridges," of a fingerprint. (Microscopes are not usually used to study fingerprints—the magnification (3x to 5x) of magnifying lenses is ideal.)

At this learning station, students use pencils and transparent tape to "lift" one of their fingerprints, and place it in their student booklet. They then study their prints with magnifying lenses, just as the experts do. Students find and label examples of typical **ridge details** like "short ridges," "forks," and "bridges." Finally, students use their newfound skills to examine two mystery fingerprints to determine if they match exactly. No microscopes are used at this station.

Please note that this learning station activity is NOT about classification of fingerprints into categories like arch, loop, and whorl. Rather, students focus on single, tiny details of the lines, or "ridges" on fingerprints. (For the study of fingerprint classification, we recommend that you present the GEMS *Fingerprinting* guide, either before or after the festival.)

While pre-teaching about fingerprints is not necessary, students will greatly benefit from experience with studying the lines and patterns on fingerprints before encountering this learning station during the festival. You might want to duplicate the station sign on an overhead transparency, and go over the various ridge details on the sample print with the whole class before the festival.

What You Need

- ❏ 4–6 hand lenses
- ❏ 4–6 sharpened No. 2 pencils
- ❏ 3 dispensers with ¾" Magic™ tape
- ❏ several pieces of white scratch paper
- ❏ damp paper towels to clean hands
- ❏ Fingerprint Ridges station sign
- ❏ Volunteer Sheet
- ❏ trash container
- ❏ (*optional*) overhead transparency of Fingerprint Ridges station sign

Getting Ready

Dampen some paper towels. Place all the materials on the table, including the station sign, and have the trash container nearby.

Going Further

1. Have students compare their fingerprints with those of other family members. Do they have any similar ridge patterns?

2. Consider presenting the *Fingerprinting* GEMS guide, have students categorize all of their fingerprints, and then solve a mystery using the standard fingerprint classifications. Then consider presenting the GEMS guide *Mystery Festival* which combines many forensic tests, including fingerprinting, into a compelling mystery format for both younger and older students.

3. Have students investigate the use of fingerprints in forensics. Invite a member of your local police department to come and demonstrate how fingerprints are taken from a crime scene and describe how they are identified using magnifying lenses. Ask the police representative how and in what ways they might take fingerprint **ridges** into account in the course of their investigations.

4. Consider presenting activities from *The Private Eye* curriculum which is listed in the "Resources" section.

"...*The texture of Cells of Cork and of some other frothy Bodies could not be so curious, but that possible, if I could use some further diligence, I might find it to be discernable with a Microscope... me thinks, it seems very probable, that Nature has in these passages, as well as in those of Animal bodies, very many appropriated Instruments and contrivances, whereby to bring her designs and end to pass, which not improbable, but that some diligent Observer, if helped by Microscopes, may in time detect.*"

Robert Hooke, Micrographia, 1665

Activity **2**
Fingerprint Ridges

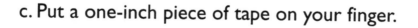

What to Do

1. Look at your fingers with a hand lens. Notice any small details.

2. Take your fingerprint (any finger or thumb is fine):

   a. Scribble on scratch paper with a pencil.

   b. Rub your finger on the pencil scribbles. Be sure to get the front and sides of the finger blackened, not just the tip.

   c. Put a one-inch piece of tape on your finger.

   d. Pull off the tape, and stick it on your booklet.

3. Look closely at the ridges (little black lines) with a hand lens.

Activity **2**

Fingerprint Ridges

Questions

A. Do you have any ridges like the ones below? Draw lines to label some of your ridges.

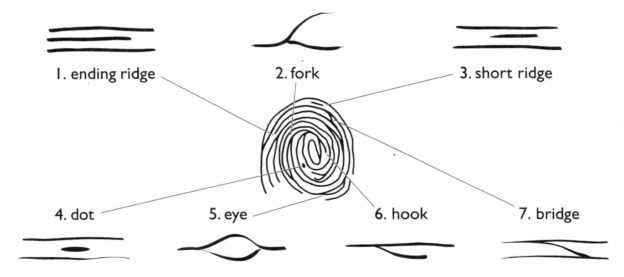

1. ending ridge   2. fork   3. short ridge

4. dot   5. eye   6. hook   7. bridge

B. Compare the ridge details in these two fingerprints. Circle any differences you see.

Print A   Print B

Volunteer Sheet, SIDE ONE

Activity 2: Fingerprint Ridges

Tips for Managing the Station

1. As students use hand lenses to examine their hands, thumbs, and fingers, encourage them to share any discoveries, whether they're fingerprint patterns, scars, scabs, or just dirt.

2. Students may need some help lifting and taping prints. It doesn't matter which finger or thumb they choose. Remind students to blacken the front and sides of the finger, not just the tip. Tell them to use a short piece of tape, about an inch and a half. Encourage them to tape a few practice prints on scratch paper until they get a clear print. Remind them to wipe the graphite off after each try, to avoid smudging the next print.

3. After students have had a chance to freely observe their print with the hand lens, guide them in noticing the black lines, or ridges. Encourage general observations and comparisons about the ridge shapes, like "hills" or "circles." Then **help students understand the goal** of this activity, **to look at one ridge at a time, and study the way it is shaped.** Use the blown-up print on the station sign to help them see what they are looking for in the context of a whole print.

4. The seven ridge details on the sign are listed in order from most to least commonly found. Tell students not to expect to find all or even most of these ridge details on their own prints. It may be best for most students to start by looking for the first three types of ridge details. Suggest that they use the tip of their pencil as a pointer to help them focus on details of their print. If they are not sure about what they are seeing, have them ask a friend to take a look. Scientists often ask other scientists to double-check their discoveries.

5. The two Mystery Prints—Prints A and B—in their booklets are very similar. Challenge students to find any differences they can and circle the differences in their booklets. Finding even one is fine.

6. Tidy up the station by throwing out heavily used scratch paper and used paper towels.

In commendation of ye microscope

Of all the Inventions none there is Surpasses
ye noble Florentine's Dioptrick Glasses
For what a better, fitter guift Could bee
in this World's Aged Luciosity.
To help our Blindnesse so as to devize
a paire of new & Artifical eyes
By whose augmenting power wee now see more
than all the world Has ever dounn before

Henry Powers, 1664
(in the original old English)

Activity 3: Dots and Dollars

Overview

At this station, students investigate a range of images to determine how they are formed. Photographs are made in different ways. Some are printed with a chemical process that keeps colors true. Postage stamps are printed in a variety of ways, and may be made up of dots or lines or other elements. Money is printed using a series of lines and spaces, and there are many "hidden" messages and pictures designed to foil counterfeiting attempts. See how many of them your students can find.

After exploring a variety of images, students focus on colored newspaper pictures and discover how dots of only a few colors can be combined to give the appearance of a range of other colors. Almost all color photographs in newspapers, magazines, and books are printed using various combinations of tiny dots of only three colors (magenta, cyan, and yellow) with black added (see "Behind the Scenes" for more information). As students look at these images under the microscope, they can see how these images are formed. The placement of the dots is critical, resulting in the correct colors as well as the outline of a clear image. You may have seen a poorly printed picture with areas of one color overlapping with the others, blurring the image.

What You Need

- ❏ at least two microscopes
- ❏ extension cords, power strips, and extra lamps, if necessary
- ❏ 4–6 hand lenses
- ❏ 1 piece of a cut up paper dollar bill
- ❏ 1 postage stamp
- ❏ 1 piece of film negative
- ❏ a few black and white newspaper photos
- ❏ 6 or more colored newspaper photos or illustrations
- ❏ a few magazine pictures
- ❏ a few business cards
- ❏ 1 penny
- ❏ Dots and Dollars station sign
- ❏ Volunteer Sheet
- ❏ (optional) movie film, Georges Seurat picture (See Going Further #3, below). You and your students could also try other prints, lithographs, or interesting papers and stationery.
- ❏ (optional) an example of a colored newspaper photograph that is poorly printed for students to observe and compare dot placement to clearer photos

Getting Ready

Place all materials on the table and set up the station sign. Try focusing on several of the samples with microscopes. To see the surfaces of more opaque or thicker items, you may need to place a lamp so it shines on top of the object rather than up through the bottom of the microscope stage.

Going Further

1. Have students find out more about how the U.S. Treasury Bureau of Engraving and Printing makes money and guards against counterfeiting. Bills printed after 1990, in denominations of $10 or larger, have the most interesting microscopic details.

2. At a bank with a foreign currency exchange, or a coin collector's store, get samples of foreign paper money. Colored, microscopically-interesting paper currency is sometimes available in small denominations.

3. The painter Georges Seurat (1859–1891) founded the 19th century French school of Neo-Impressionism known as pointillism, which used small dots or strokes of color to create images and light patterns. He created huge canvases in which individual dots are too small to be seen when viewed from a distance, but are obvious with close inspection. Have students investigate the art of impressionism and pointillism, and make dot pictures of their own.

4. Consider presenting the GEMS guide *Color Analyzers*, in which students utilize diffraction grating, and both green and red filters, to create "secret messages" and explore the visible spectrum to learn about light and color. This guide would make an excellent and challenging accompaniment to this station, because it requires a fairly complex thinking process to understand the color printing process from the standpoint of the basic concepts conveyed in *Color Analyzers*.

5. Teach an art lesson related to primary colors, or, for older students, consider learning more about how printing creates full-color from a four-color process. It is important to clarify that the "primary colors" in mixing paints are not the same as those also referred to as primary colors in the printing process. Taking a field trip to a nearby printer or having a representative visit the class could also lead to a deeper investigation of several related subjects, including half-tones and differing resolutions. Some of these issues also apply to computer printing processes.

6. Encourage students to read and write short essays on the book by Ruth Heller entitled *Color* that graphically depicts the color printing process. See "Literature Connections" for the full listing.

7. Have students draw a penny: (a) from memory, before observing; (b) from memory, after observing; and (c) while observing. Compare!

Dots and Dollars

What to Do

1. Look at all the pictures, newspaper, stamps, money, and other items:

- with just your eyes

- with a hand lens

- with a microscope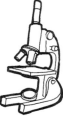

2. Look carefully at a colored newspaper picture:

- with just your eyes

- with a hand lens

- with a microscope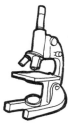

Dots and Dollars

Questions

A. What do you notice when you look at all the items with a hand lens?

B. What do you notice when you look at the items with a microscope?

C. When looking at the colored newspaper:

• What do you notice about the dots?

• Are all the dots the same size? Are they the same distance apart?

• How many different colors of dots can you find? What colors are they?

Volunteer Sheet, SIDE ONE

Activity 3: Dots and Dollars

Tips for Managing the Station

1. Because the pieces of paper are small, it is easy for them to fall on the floor. Be sure to watch for them and have students pick them up.

2. Encourage students to look closely and notice the differences between the samples. They may take a quick look and think that there is nothing to see beyond the obvious.

3. If students are having trouble viewing an object with the microscope because it is too thick, have them move the light so it shines on top of the object rather than up through the bottom of the microscope stage. By having the light come from above, the students should be able to see the markings on the paper and other objects more clearly.

"In the age of One World, the power of the microscope will be one doesn't know how many times greater than that of [the instrument of] today. [Viewed through the instrument of today] an ant looks like an elephant. [Viewed through the instrument of] the future, the size of a microbe will be like that of the great, skyborne p'eng bird."

K'ang Yu-wei (1858-1927)
Ta T'ung Shu:
The One-world Philosophy of K'ang Yu-wei
translated by L.G. Thompson, 1958

Activity 4: Fabrics

Overview

At this station, students observe a variety of fabrics, using hand lenses and microscopes, then choose one fabric to study carefully to try to determine how it was made. Fabrics are made in three main ways. Weaving is accomplished by the use of cross-threads through which other threads are passed in an alternating pattern. Woven fabrics have straight lines at right angles when viewed with a microscope. Knitted fabrics are made from one continuous thread or strand of fiber that is looped around itself. These fabrics have curved, interlocking loops when viewed with a microscope. Pressed fabrics are made much the same way as paper, by adhering clumps of fibers together. These fabrics look like a random assortment of fibers when viewed with a microscope.

What You Need

- ❑ at least two microscopes
- ❑ extension cords, power strips, and extra lamps, if necessary
- ❑ 4–6 hand lenses
- ❑ 6–12 small fabric samples (include woven, knitted, and pressed)
- ❑ Fabrics station sign
- ❑ Volunteer Sheet
- ❑ *(optional)* several small pieces of Velcro, both sides

Note: *Some woven fabrics are made of all different colors of threads woven together to make a design. Others are made of all white threads woven together, after which color is printed onto them. If possible, get samples of both kinds. Also, try to include some less densely-woven fabrics, such as burlap, muslin, fish net fabric, and/or pantyhose, since the weave will be more easily observed. In general, we've found that the weave or knit of **light-colored fabrics** tends to be more readily apparent under the microscope.*

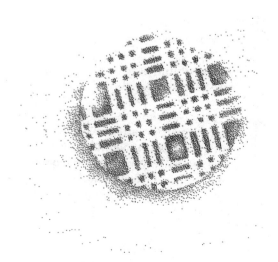

Getting Ready

1. Place fabric samples and hand lenses on the table with the microscopes. If you have obtained pieces of Velcro, set those out too. Set out the station sign.

2. Try focusing on several of the fabrics with microscopes. To see the surfaces of more opaque or thicker ones, you may need to place a lamp so it shines on top of the fabric rather than up through the bottom of the microscope stage.

Going Further

1. Have students make pipe cleaner models of woven and knitted fabrics.

2. Ask students to draw and describe a machine that could make each type of fabric.

3. Have a collaborative weaving project. Prepare a "loom" by threading yarn across a shallow cardboard box. Cut slits about an inch deep and an inch apart along two opposite sides of the box, and thread yarn back and forth. Have students collaborate to "weave" a piece of fabric: Each student wraps one end of a small pipe cleaner around a piece of yarn, to act as a "needle." Have them use the pipe cleaner to weave the strand into the yarn on the loom. They unwrap the pipe cleaner when finished, and press their thread against the other students' threads. A comb will be helpful.

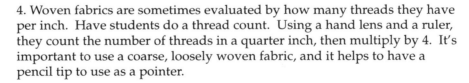

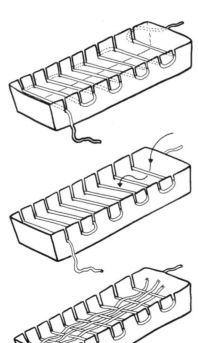

4. Woven fabrics are sometimes evaluated by how many threads they have per inch. Have students do a thread count. Using a hand lens and a ruler, they count the number of threads in a quarter inch, then multiply by 4. It's important to use a coarse, loosely woven fabric, and it helps to have a pencil tip to use as a pointer.

5. Have students investigate how color is added to fabrics. Are the individual threads themselves a certain color before they are woven or knitted, or is the color applied after the fabric is formed?

6. Have students do research and compare natural and synthetic fibers used in making fabrics.

7. Have students investigate Velcro with a microscope and try to determine how it works and how it is made. Students could also do research on the interesting story behind how a Swiss scientist, George de Mestral, invented Velcro. After a walk in the woods in 1948, he found seedpods sticking to his clothes. He examined their tiny hooks under a microscope, then spent the next eight years figuring out how humans could adapt the same idea to create Velcro.

Activity 4
Fabrics

What to Do

1. Look at all the fabric samples.

2. Pick one fabric. Draw and describe it when you look:

- with just your eyes

- with a hand lens

- with a microscope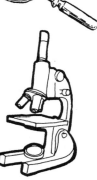

Activity **4**
Fabrics

Questions

Do you think your fabric is woven, knitted, or pressed? Why do you think so?

woven

knitted

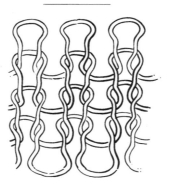

pressed

Activity 4: Fabrics

Tips for Managing the Station

1. Be sure that the small pieces of fabrics do not fall on the ground. If they do, have students pick them up.

2. As students examine the various fabric samples, ask them to describe what they see. You might also encourage them to sort the samples into piles according to how they are constructed: woven, knitted, or pressed. Sometimes it is interesting to look at both sides of a sample.

3. If students are having trouble viewing the fabrics with the microscope because they are too thick, have them move the light so it shines on top of the fabric. By having the light come from above, the students should be able to see the texture of the fabric more clearly.

"...It is because simplicity and vastness are both beautiful that we seek by preference simple facts and vast facts; that we take delight, now in scrutinizing with a microscope that prodigious smallness which is also a vastness..."

Henri Poincaré (French mathematician, late 1800s)

Activity 5: Salts

Overview

At this station, students observe and compare crystals of various salts. They identify the salt in a make-believe highway spill. They then make a crystal "suncatcher" by coloring a clear plastic lid with crayons and then adding a small amount of a solution of Epsom salts. As the solution evaporates, students use hand lenses to watch crystals form on the surface of the lid. Students always enjoy getting to take home their suncatcher.

Chemical solids that form with regular shapes are called crystals. In different chemicals, the molecules bond and line up in different arrangements, based on the charges of the atoms within them, leading to the formation of crystals with different shapes. In this case, the Epsom salt crystals from the box have the form of elongated oval crystals. But when these crystals are dissolved in hot water, after the water evaporates and the crystals form again, the crystal pattern is different. Now, the crystals tend to make a star-like pattern. This phenomena is taken advantage of at this station to make the crystal suncatchers, which students can take home with them.

Please Note: **The Crystal Suncatcher activity is usually one of the most popular festival activities for students. However, some teachers of older students may find making suncatchers too elementary, and choose to present only the main station activity.**

What You Need (for the Salt Crystals activity)

- ☐ at least two microscopes
- ☐ extension cords, power strips, and extra lamps, if necessary
- ☐ 4–6 hand lenses
- ☐ 10 labeled hole-punch slides (two for each salt)
- ☐ about a tablespoon of each: table salt, rock salt, alum, boric acid, Epsom salt (local grocery stores and pharmacies should carry these)
- ☐ 5 plastic ziplock sandwich bags (for shaking hole-punch slides in salts)
- ☐ several sheets of black construction paper
- ☐ Salts station sign
- ☐ Volunteer Sheet

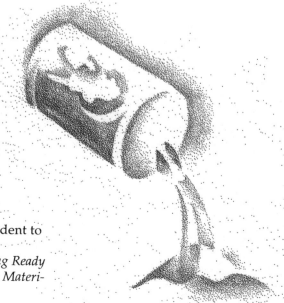

What You Need (for the Crystal Suncatcher activity)

- ☐ 1 package Epsom salt
- ☐ clear, plastic deli-type container lids (enough for each student to have one)
  See additional information on obtaining deli lids in the "Getting Ready One month before the festival" on page 18, and in "Sources for Materials," page 121.

❏ 1 small squeeze bottle of liquid dishwashing detergent, any brand
Each student smears a drop of dishwashing detergent on the lid before the volunteer adds the salt solution. The soap allows the small amount of salt solution to spread out on the lid more easily. (The soap reduces the surface tension of the water, so it doesn't bead up on the plastic lid.)

❏ 1 hot pot, 2 cup size or larger
❏ 1 metal teaspoon
❏ crayons (or permanent felt-tipped colored markers)
Crayons are the best alternative for decorating the suncatchers, even though they don't produce bright colors on the plastic lids. Water-based felt markers don't work on the plastic lids. Permanent felt markers do make more vivid colors on the suncatchers than crayons, but can be problematic if vandalism and/or graffiti are of local concern. Permanent markers also may have a solvent smell. If you do use permanent markers, have the volunteer monitor them to make sure they stay at the learning station.

❏ paper towels
❏ Crystal Suncatcher station sign
❏ Volunteer Sheet

Getting Ready

1. If possible, set up two tables (or two ends of a long table or counter) for this two-part learning station. (It doesn't matter which area of the station students visit first.) The area for the suncatcher activity needs to be near an electrical outlet.

2. **An adult volunteer must oversee the crystal suncatcher activity at all times.** You may want to arrange for a second volunteer in the area where students observe five salt crystals with hand lenses and microscopes.

Getting Ready (for the Salt Crystals activity)

1. Prepare two hole-punch slides of each of the five salts (table salt, rock salt, alum, boric acid, and Epsom salt). Directions for making slides are on page 21. Because the rock salt crystals are large, place an individual crystal of rock salt on the slide. Keep some extra rock salt handy in case the crystal falls off the slide during the activity.

2. Set out the microscopes, some of the hand lenses, ten hole-punch slides, and the Salts station sign. Set out pieces of black paper for students to use as a contrasting background when observing salt crystals with hand lenses.

Getting Ready (for the Crystal Suncatcher activity)

1. Fill the hot pot with a mixture of half water, half Epsom salt. Plug in the hot pot until it is boiling, then unplug it, but keep it hot.

2. Be sure the electrical cord for the hot pot is out of the way and/or taped down.

3. Place crayons, detergent, spoon, the rest of the magnifying lenses, and paper towels in this area. Set out six lids on paper towels, and keep the other lids handy.

4. Set out the Crystal Suncatcher station sign.

5. If there is time before the festival, have the volunteer familiarize him/herself with the procedure by making a practice suncatcher.

Going Further

1. Present the GEMS unit *Stories in Stone* and explore crystal formation in the context of geology and the earth sciences.

2. Explore the geometry of crystal shapes further with the GEMS unit *Build It! Festival* and/or other geometry units you use in teaching mathematics.

3. Have students investigate the formation of crystals with the other salts at the station, such as rock salt, alum, and boric acid. Do they resemble the original crystals?

4. Have students investigate the concentrations of salts that make various types of crystals. How concentrated does the solution have to be before crystals will form? Does the concentration affect how the crystals look?

5. Have students watch the Epsom salt crystals form under a microscope in a small drop on a well slide or on a clear tape slide (such as the ones used in Activity 1, Up Close).

6. Consider ordering the video "Crystals: They're Habit Forming" which is listed in the "Resources" section.

"It is rather remarkable how slow American chemists have been in realizing the importance of the microscope as an adjunct to every chemical laboratory... (The microscope is) as much a necessity in every analytical laboratory as is the balance."

E. M. Chamot
Journal of Applied Microscopy 2: 502 (1899)

Activity 5
Salts

What to Do

With a hand lens or microscope, look at five different kinds of salt:

- Epsom salt

- table salt

- alum

- rock salt

- boric acid

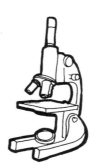

Activity **5**
Salts

Questions

A. What does each kind of salt crystal look like? Draw and describe all five kinds of salt crystals. What shapes are they? Sizes? Colors?

B. A truck has spilled some kind of salt on the highway.

The crystals look like this:

What kind of salt do you think it is? Why do you think so?

Activity **5**, Part 2

Make a
Salt Crystal Suncatcher

What to Do

1. Take a plastic lid.

2. Color the outside in any design you want.

Outside   Inside

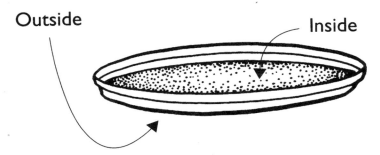

3. Spread one drop of detergent inside the lid with your finger.

4. Ask the volunteer to put a spoonful or two of hot Epsom salt solution in the lid.

5. Swish the liquid around until the lid is completely covered.

6. Let it dry without touching it.

7. Watch crystals form inside the plastic lid. Observe with a hand lens.

66   *Activity 5*

Volunteer Sheet, SIDE ONE

Activity 5: Salts

Tips for Managing the Station

If you are the only adult at the Salts learning station, your responsibility is to supervise the hot pot at the suncatcher activity.

1. Students may wish to make a suncatcher before they look at the crystals. That's okay, but if the suncatcher area is too crowded, have them do the Salts activity first.

2. Students may crowd around the suncatcher table—it's one of the most popular activities. Be firm in enforcing the "maximum of six students at a time" rule. Put only a few lids out at a time, replacing them when a new group of students comes. Be sure each student makes only one suncatcher.

3. Double-check that the hot pot cord is taped down and away from students' feet.

4. For the suncatcher activity, *the "inside" is the side with a shallow depression surrounded by a rim.* (See station sign.) When students have colored the "outside" of the plastic lid and covered the inside with a drop of liquid detergent, your job is to add one to two teaspoons of salt solution to each lid. There should be just enough solution to cover the lid when swirled around. **Add the smallest amount you can that will cover the lid.** Adding more liquid results in a longer wait for the crystals to form.

(Students may ask about adding the drop of detergent. The detergent breaks the surface tension of the salt solution and this prevents the solution from "beading up" on the plastic. So, the detergent helps the salt solution spread out more evenly on the lid.)

5. The solution of half Epsom salt and half water should be kept hot (because more salt dissolves in hot water than cold), but does not have to be boiling.

6. Once they have watched their own crystals forming, encourage students to compare the crystals that form on their suncatchers to the Epsom salt crystals on the hold-punch slides. (If students ask why the shape of the Epsom salt crystals is different in their suncatchers, you may want to explain that heating the salt in water results in a changed crystal shape.)

7. Salt water, microscopes, and electric power cords are a bad combination. Please be sure to have students wipe up spills immediately.

Activity 6: Sand

Overview

At this station, students compare sand samples from several locations based on the color, size, and shape of the sand grains. They then locate and mark the source of their sand on a world map. Look into enough handfuls of sand, and you will see nearly every color imaginable. Sand is made of crushed rocks, shells, bones, glass, metals, and/or corals, depending on location. Sand is produced as the result of waves crashing, winds blowing, and streams tumbling—natural processes of erosion that eventually make nearly every solid substance into sand. Each sand sample has its own story, and clues to its origins can be found in the colors, shapes, and sizes of its grains. Colors give a clue as to the materials sand is composed of; shapes tell a bit about the age of the sand—usually, the older the sand grain, the smoother the edges. The size of the grains indicate the type of beach the sand is from—larger grains are usually from areas with vigorous wave action, while smaller grains can settle from the slower, calmer waters of secluded bays and estuaries.

What You Need

- ❏ at least two microscopes
- ❏ extension cords, power strips, and extra lamps, if necessary
- ❏ 4–6 hand lenses
- ❏ 3–6 sand samples from several locations, 1 tablespoon each (see note below)
- ❏ 3–6 ziplock sandwich bags (to store sand samples)
- ❏ 6–12 labeled hole-punch slides (two of each of the sand samples)
- ❏ world map (or globe)
- ❏ small Post-it® Notes or adhesive dots to mark locations on map
- ❏ Sand station sign
- ❏ Volunteer Sheet
- ❏ (optional) some rock, shell, and coral samples

Note: While it is nice to have sands from a variety of geographical locations around the county or the world, it is not necessary! Several samples from different parts of the same town can be quite varied in size, texture, color, etc. Please see page 121 for ideas on collecting sand samples and information on the International Sand Collector's Society.

Getting Ready

1. Prepare two hole-punch slides of each sand sample. Labels should tell where the sand is from. For directions on preparing hole-punch slides, see page 21.

2. Set out slides, hand lenses, microscopes, map, and Sand station sign.

Going Further

1. Have students collect sand samples from around the world. See if they can create a network of "sand pals" with whom to exchange sands.

2. For younger students, the GEMS guide *On Sandy Shores* is a nice accompaniment to this learning station activity. While that unit is written for Grades 2–4, many of the sand activities can be extended to higher grades. For older students, consider presenting the GEMS unit *Stories in Stone*, for investigations of rocks and minerals and the processes that create and transform them.

3. Make your own "class sand." Have students bring in rocks, shells, and other materials to contribute. The teacher then safely pounds the materials with a hammer, of course making sure to wrap the materials inside a pillowcase or towel. Students could also try making their own sand and others could guess what materials are in it.

4. Make "candy sand" by shaking an assortment of hard candies in a plastic container. Have students challenge each other to identify which candies they used to make the sand.

"To see a world in a grain of sand

and a heaven in a wild flower,

hold infinity in the palm of your hand

and eternity in an hour."

William Blake

Activity **6**
Sand

What to Do

1. Look at all the sand samples:

- with a hand lens

- with a microscope

2. Look for the sand with:

- the most different colors

- the largest grains

- the smallest grains

- the sharpest edges

- the roundest edges

Activity **6**
Sand

Questions

A. Pick one type of sand. Describe its color, grain size and shape, and anything else about it.

B. Estimate how many grains of sand are on the hole-punch slide.

C. Where is your sand from? Label it on the map.

Volunteer Sheet, Side One

Activity 6: Sand

Tips for Managing the Station

1. If students are having trouble viewing their sand grains, have them move the light so it shines on top of the sand rather than up through the bottom of the microscope stage. This way, students should be able to see the individual grains of sand more clearly.

2. If students need help finding where the sands came from on the map, give them hints about the general area on the map where they should look.

Activity 7: Kitchen Powders

Overview

At this station, students examine several white powders they might find in their own kitchens. They are challenged to compare the shape, color, and size of the grains of each individual powder. Using these observations, they then attempt to identify a mystery powder and a mystery mixture of two powders. All of these kitchen powders should be familiar to the students, yet they may never have looked closely at any of them. These white powdery substances, which might at first glance look the same, turn out, upon close examination, to have distinctive appearances. Some are crystals, or regularly-shaped grains, while others are irregular shapes or small flakes of material. Some are translucent or transparent, while others are opaque. Although all can be described as white, small differences in color are apparent to the careful observer.

As your students examine and describe these characteristics, and use them to distinguish between substances, they build powerful skills of observation and description. They develop vocabulary skills as they find, and write down in their booklets, words that precisely distinguish between these similar substances.

What You Need

- ❏ at least two microscopes
- ❏ extension cords, power strips, and extra lamps, if necessary
- ❏ 4–6 hand lenses
- ❏ 18 hole-punch slides (See "Getting Ready," below)
- ❏ 1 or 2 spoonfuls of each of 7 powders: table salt, white granulated sugar, baking soda, cornstarch, cream of tartar, laundry detergent, white flour
- ❏ 9 ziplock sandwich bags (for shaking hole-punch slides in powders)
- ❏ Kitchen Powders station sign
- ❏ Volunteer Sheet

Getting Ready

1. Prepare two labeled hole-punch slides of each of the seven powders. (Directions for making slides are on page 21.) Make two extra slides of sugar, and label them "Mystery Powder." Make two more slides of a mixture of salt and detergent, and label them "Mystery Mixture."

2. Place all 18 slides on the table, along with the microscopes and hand lenses.

3. Set out the Kitchen Powders station sign.

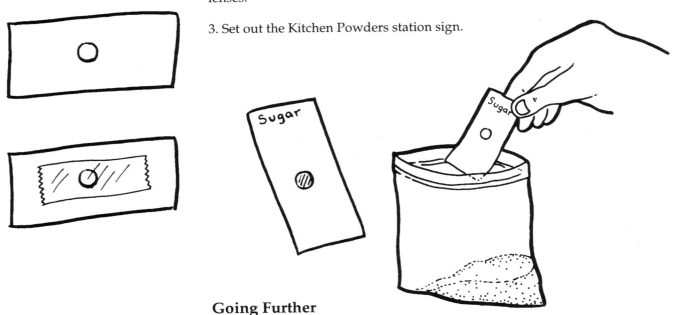

Going Further

1. Have students make hole-punch slides of other "Mystery Mixtures" and test each other in their ability to identify these substances. Are there any that are impossible to distinguish?

2. Have students investigate other kitchen powders, such as citric acid, and sugar substitutes. Make sure they know that they should use safe powders such as foods, *but not any medicinal or cleaning /scouring powders.*

3. Challenge students to find other ways to distinguish between these substances, such as testing the pH, using an iodine test, or watching the reactions with water or vinegar or other liquids. Have them check with you first to ensure that safety precautions are taken as needed. Encourage them to draw on their other experiences in testing materials.

Kitchen Powders

What to Do

Study each powder with a hand lens and microscope. Notice the shapes, colors, and sizes of the grains.

- table salt

- cream of tartar

- laundry detergent

- baking soda

- cornstarch

- white granulated sugar

- white flour

Kitchen Powders

Questions

A. Pick any two powders. Write the names of the two powders in your student booklet, and carefully draw and describe each of them.

B. What is the mystery powder? How do you know?

C. What is the mystery mixture? How do you know?

Activity 7: Kitchen Powders

Tips for Managing the Station

1. Try to be sure students inspect all seven different powders, and encourage them to notice shape, size and color, and any other details.

2. If students are having difficulty viewing the slides because the powders are thick, have them adjust the light so it shines on top of the slide, rather than only up through the bottom of the microscope stage. By having the light come from above, the students should be able to see the characteristics of the powders more clearly.

3. Remind students to write down their descriptions of two of the powders in their student booklets and draw them. Encourage students to use descriptive language by asking questions like, "What does it remind you of?" "What shape, size, texture is it?" If time and interest allow, encourage students to compare and describe more than two powders.

Do not despise the creatures
because they are minute. . .
doubt not that in these tiny
creatures are mysteries more
than we can ever fathom.

Charles Kingley
Glaucus or the Wonders of the Shore, 1855

Activity 8: Small Creatures

Overview

It's probably safe to say that creatures generally referred to as "bugs" are what most students would like to look at first when using a microscope or hand lens. At this station, students get to examine several such familiar small creatures—insects, spiders, and even "true bugs." Through magnification, even the most common and familiar crawling or flying "bug" can be appreciated for the complex, fascinating organism it is. This is an especially high-interest learning station, and another marvelous way to appreciate the microscope as a tool for scientific investigation.

Please note: In common usage, the word "bugs" is used to refer to many small creatures. In scientific terminology, however, this is not accurate. "True bugs" are a subset—order Hemiptera—of insects. The milkweed bug is one example. Therefore, technically speaking, all other insects, as well as spiders and isopods, are not "bugs." Insects, spiders, and isopods (classes Insecta, Arachnida, and Isopoda) are subsets of arthropods (phylum Arthropoda) which also includes crabs and lobsters of the class Crustacea—as well as brine shrimp.

What You Need

- ❏ at least two microscopes (having at least one dissecting microscope is desirable, but a good-quality compound microscope set on the lowest magnification also works well)
- ❏ extension cords, power strips, and extra lamps, if necessary
- ❏ 4–6 hand lenses
- ❏ 6–10 hole-punch slides
- ❏ 6–10 dead insects and spiders, or parts, such as wings, antennae, etc.
- ❏ a few 3" x 5" index cards
- ❏ a few toothpicks with flat end
- ❏ a few empty yogurt containers or film canisters
- ❏ a box, tray, or paper plate for storing insects
- ❏ one pair of tweezers
- ❏ Small Creatures station sign
- ❏ Volunteer Sheet
- ❏ (*optional*) a scanning electron micrograph of a fly head, or other interesting picture relating to insects

Getting Ready

1. See "Getting Ready One month before the festival" (page 18) for information about gathering small creatures, and "Getting Ready One week before the festival" (page 22) for more on enlarging the opening of the hole-punch slides.

2. Mount the six to 10 dead specimens, gathered by student volunteers, on hole-punch slides. (Some teachers like to have even more than 10 specimens. The variety is great, but be aware that having a huge collection can be hard to handle at the station.) For larger specimens, make the opening in the slide larger by punching two or three holes in a row, and covering them with tape. Be careful not to touch the organism itself. Using toothpicks and index card, or tweezers, carefully place the organism on the sticky side of the tape—on its back, legs sticking up, if possible. If you know what the specimen is, label the slide. These specimens will last longer if they're mounted in taped-shut plastic containers (small Petri dishes, hardware bubble-packs, etc.).

3. Slides with these kinds of specimens are extremely fragile. A shoebox or plastic box with a lid would be great for storing your slides, but if you don't have one, use a paper plate or cafeteria tray. If some specimens break apart, save interesting parts like wings, heads, and legs to make extra slides.

4. Set out microscopes, hand lenses, prepared slides, and the Small Creatures station sign.

Going Further

1. Tie this activity in with the GEMS guides *Terrarium Habitats*, which includes the study of isopods, and/or *Schoolyard Ecology*, which features activities on ants and spiders.

2. Start a terrarium or meal worm colony. Compare the larvae, pupae, and adults as they go through their life cycle.

3. Investigate bees. Visit a beekeeper and find out about how bees pollinate crops and make honey. For younger students, consider presenting the GEMS guide *Buzzing A Hive*.

4. Build spider webs with yarn. Compare the different types of webs that different kinds of spiders make. See the GEMS guide *Schoolyard Ecology* for a web identification key and other information on spiders.

Activity **8**

Small Creatures

What to Do

1. Look at several different creatures with the hand lens and microscope.

2. Choose one creature. Draw the whole creature, as seen through a hand lens. If you know the name of the creature, label your drawing.

3. Use the microscope to look at one part of your creature. Describe and draw the part in detail.

Activity **8**
Small Creatures

Questions

A. How many legs does it have? What do the legs look like?

B. Does it have wings? What do they look like?

C. What do the eyes look like? The mouth?

D. What other body parts do you see? Describe them.

Volunteer **S**heet, SIDE ONE

Activity 8: Small Creatures

Tips for Managing the Station

1. For this station, a dissecting microscope is great. If you have only compound microscopes, be sure to use the lowest power. Have students move a table lamp so light shines on top of the creature as well as up through the bottom of the microscope stage. This should help them see details of the specimen more clearly.

2. If a large specimen is hard to see with the microscope, have students use a hand lens, or try using the microscope for a small creature or just a wing.

3. Ask students to describe details of the body parts they see and to think about what the various structures might be designed to do.

4. Remind students not to touch the specimens directly. If they do so accidentally, you may want to suggest they wash their hands after the festival.

5. Some students may think it's amusing to startle others who are nervous around even dead insects/spiders. Be on the lookout for such behavior, encourage respectful inquiry at the station, and caution students to be careful with the specimens.

In ordinary speech, people often call all such small creatures—"bugs." Scientifically speaking, however, this is not accurate. There are some organisms that are classified as true bugs, such as the milkweed bug. But many other organisms people call "bugs" are part of a larger group of insects. Insects, along with spiders and isopods, are part of an even larger grouping or phylum, called Arthropoda, that also includes some marine organisms, such as brine shrimp, crabs, and lobsters.

"One can be fooled by appearances, which happens only too frequently, whether one uses a microscope or not."

Voltaire, Micromegas

Activity 9: Brine Shrimp

Overview

At this station, students observe live brine shrimp using a hand lens and microscope. They compare the overall appearance, structures, and movement patterns of adults, larvae, and eggs. They learn to identify and distinguish between male and female adult brine shrimp. They also use their math skills to estimate the number of eggs on a slide.

If your students have ever heard of "sea monkeys," then they know what brine shrimp are. Brine shrimp, scientific name *Artemia*, are small crustaceans that live in the salt ponds in bays and estuaries, providing food for the numerous birds that use these wetlands as breeding and feeding grounds. Although brine shrimp are large enough to see with the naked eye, using a microscope allows students to see the legs, head, eyes, and gills, and pick out the females, with egg sacs at the base of their tails, and the males, with horns on both sides of their heads.

Brine shrimp eggs can remain viable for years if stored in a cool, dry place (a refrigerator is ideal). All it takes for them to come to life is a little warm, aerated salty water, and with some luck, the larvae are swimming around within a few days. Larvae look different from the adults, being basically a small triangle with a few legs. In the wild, the larvae become adults in about eight days, and adults then live an average of 50 days. In an aquarium, however, adults usually last only a few days, and rarely will larvae survive into adulthood.

What You Need

- ❏ at least two microscopes (having at least one dissecting microscope is desirable, but a good-quality compound microscope set on the lowest magnification also works well)
- ❏ extension cords, power strips, and extra lamps, if necessary
- ❏ 4–6 hand lenses
- ❏ 4–6 plastic depression slides
- ❏ 6 hole-punch slides of brine shrimp eggs
- ❏ 2 cottage cheese type containers
- ❏ 2 medicine droppers
- ❏ 1 container brine shrimp eggs
- ❏ 3 tablespoons salt (kosher is best, but table salt is fine)
- ❏ 1 ounce of live brine shrimp adults
- ❏ some newly hatched brine shrimp larvae
- ❏ 1 sheet 8 ½" x 11" paper
- ❏ paper towels
- ❏ Brine Shrimp station sign
- ❏ Volunteer Sheet
- ❏ trash container

Getting Ready

The key to success for this station is planning ahead. The "Getting Ready" section for the whole festival, pages 16–24, includes each step, from locating sources of live brine shrimp a month before the festival to hatching eggs two or three days before the festival.

1. Place one cottage cheese type container of brine shrimp adults and one container of newly hatched larvae at the station. Label the containers "Brine Shrimp Adults" and "Brine Shrimp Larvae" and place a medicine dropper and some depression slides beside each.

Please note: **If, for some reason, your brine shrimp eggs did not hatch, set up the station without the larvae, and tell students to disregard the questions about larvae on the station sign. Having brine shrimp adults and eggs will be enough to make the station a success. Also, if you have a few dead brine shrimp, keep them at the station as well; sometimes they are easier for students to observe than the live, moving ones!**

2. Set out six hole-punch slides of brine shrimp eggs (directions for making them are on page 21).

3. Write "Estimate how many brine shrimp eggs are on the slide" at the top of a sheet of paper and set it out on the table along with one of the slides of eggs. Set hand lenses, paper towels, and the station sign on the table along with the microscopes. Put a trash container near the station.

Going Further

1. Challenge your students to design an experiment to determine whether adult brine shrimp prefer dark or light, cold or warm, or gradients in other environmental conditions.

2. Investigate the conditions that lead to the most effective hatching of brine shrimp eggs. You can experiment with temperature, salinity, type of water used (rainwater, pond water, tap water), light levels, pH, etc.

Activity **9**
Brine Shrimp

What to Do

1. Use the dropper to put a few brine shrimp **adults** on a plastic slide. Look at them with a hand lens or microscope.

2. Put some **larvae** on another slide. Look at them with a hand lens or microscope.

3. With a hand lens or microscope, look at the **eggs** on the hole-punch slide.

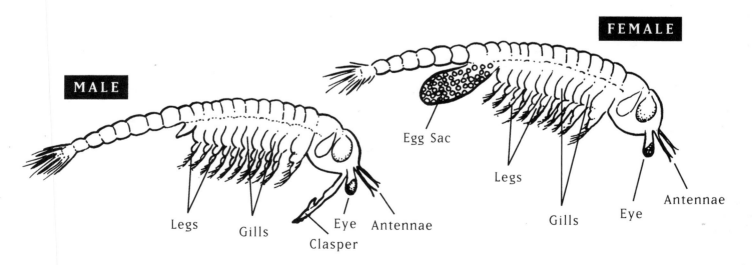

MALE

Legs   Gills   Eye   Antennae
Clasper

FEMALE

Egg Sac

Legs

Gills   Eye   Antennae

Activity 9
Brine Shrimp

Questions

A. Draw and describe an adult brine shrimp. How does it move?

B. Is the brine shrimp adult a male or a female? How do you know?

C. If you see a larva, draw and describe it. How is it different from the adult?

D. How many brine shrimp eggs do you think are on the slide? Write your estimate on the paper at the station.

Volunteer Sheet, SIDE ONE

Activity 9: Brine Shrimp

Tips for Managing the Station

1. For this station, a dissecting microscope is great. If you have only compound microscopes, be sure to use the lowest power.

2. If students need help with the medicine droppers, suggest the following:

   • Squeeze all water/air out of the dropper.
   • Lower the dropper into the water and "unsqueeze."
   • Lift the dropper, hold it over the slide, and gently squeeze.

3. Students may have trouble with brine shrimp getting stuck in the droppers. If this happens, have them pull up some water into the dropper, gently tap the brine shrimp into the water, wait for them to swim around, and then squirt them out.

4. Be sure students use the plastic depression slide with the flat side down. Brine shrimp should be placed into the small "bowl" on the side facing up.

5. If students have too many brine shrimp in the slide, have them pull some out with the medicine dropper and then add a small amount of water so the remaining brine shrimp can swim.

6. If there is too much water in the slide, have the students pull off a little with the medicine dropper so the brine shrimp are confined to a smaller space. This will allow students to see them more easily.

7. If necessary, for younger students or any students who have a very short time at the station, prepare the slides for them yourself so they can have more time to observe and record.

8. Salt water and microscopes are a bad combination. Please be sure to have students wipe up spills immediately. Make sure they wipe off the bottom of the plastic slides before they put them onto the microscope stage.

Activity 10: Pond Life

Overview

At this station, students compare the organisms found in a pond. They then use their observations and a variety of art materials to construct a model of a plant or animal they saw with the microscope. These creatures are added to a class mural and compared to other plants and animals observed by classmates.

Pond water is a favorite thing to study with a microscope, for good reason. There are few things more motivating to young students than the tiny hidden worlds in a drop of the most ordinary-looking pond water. Small shrimp-like creatures, worms, snails, larvae, and even tiny fish may zoom around right under their noses. Smaller than these are the single-celled protozoans, such as ciliates, with their alien-looking movement and forms. Plants are also unusual, ranging from the strands of translucent leaves of *Elodea* to the small paired leaves of duckweed to tiny single-celled green dots. *Daphnia*, a small crustacean related to brine shrimp, is a common inhabitant of ponds. These so-called "water fleas" have a single compound eye, and they swim using their antennae rather than their legs. Unlike brine shrimp, *Daphnia* live in freshwater ponds, so you would never find these two relatives together.

What You Need (for the first activity)

- ❏ at least two microscopes (having at least one dissecting microscope is desirable, but a good-quality compound microscope set on the lowest magnification also works well)
- ❏ extension cords, power strips, and extra lamps, if necessary
- ❏ 4–6 hand lenses
- ❏ pond water, with *Elodea, Daphnia,* and/or other aquatic animals and plants (ocean water could also be used)
- ❏ 4–6 plastic depression slides
- ❏ 4–6 medicine droppers
- ❏ paper towels
- ❏ Pond Life station sign
- ❏ Volunteer Sheet
- ❏ trash container

What You Need (for the mural activity)

- ❏ black construction paper, cut into approximately 2"–3" pieces, squares or rectangles
- ❏ assorted materials for making plant and animal models, such as: pipe cleaners, styrofoam peanuts, popsicle sticks, straws, toothpicks
- ❏ 1 large piece of butcher paper, about 3' x 4' (preferably blue)
- ❏ scissors
- ❏ clear tape
- ❏ 2 bottles white glue
- ❏ Make a Model station sign

Getting Ready

The "Getting Ready" section for the whole festival, pages 16–24, includes the earlier preparation for this station's specimens, from locating sources of pond organisms a month before the festival to gathering them two or three days before the festival.

1. Set up two areas for this learning station. In the area where students will examine pond life, place the pond and/or ocean water samples, medicine droppers, plastic depression slides, microscopes, and hand lenses. Set out paper towels and the Pond Life station sign. Put a trash container near the station. Look at the pond water samples with a microscope yourself to make sure there is something living to observe!

2. In the second area, where students construct models of the plants and animals they've observed, place tape, glue, and art materials. Set out the Make a Model station sign.

3. Cut a piece of butcher paper, preferably blue, into a large drop shape. Label it, "What lives in a drop of water?" If you will have water from both a pond and the ocean, make two drops, and label one, "What lives in a drop of pond water?" and the other, "What lives in a drop of ocean water?" Tape the drop-shaped paper(s) onto the wall behind the second area of the learning station.

Going Further

1. Have students make a detailed comparison between brine shrimp and *Daphnia*, including structures, movement patterns, feeding behavior, life cycle, and habitat.

2. Challenge students to design an imaginary creature that would live in a pond. Be sure they can explain the adaptations it has for life underwater, what it eats, where it lives, how it reproduces, how it moves, and what eats it.

Pond Life

What to Do

1. Put one or two drops of pond water on a slide.

2. Look at the pond water with a microscope or hand lens.

3. Find plants and animals in the water drop. Observe them carefully.

4. In your booklet, draw and describe one plant and one animal.

5. If you have time, describe more animals and plants.

Activity 10
Pond Life

Questions

A. What colors and shapes can you see on the plant? What other details do you see on the plant?

B. What parts of the animal do you see? Legs? Wings? Eyes? Antennae?

C. How is the water animal moving?

© 1998 by The Regents of the University of California, LHS-GEMS. *Microscopic Explorations.* **May be duplicated for classroom use.**

Make a Model

of Pond Life

What to Do

1. Pick your favorite plant or animal that you saw in the pond water, and use the art materials to make a model of it.

2. Tape your creature to the mural of a drop of water. Take a look at what the other students have made.

Volunteer Sheet, SIDE ONE

Activity 10: Pond Life

Tips for Managing the Station

1. Be sure that students look at the pond water with the microscope before they make their creature for the mural.

2. If students need help with the medicine droppers, suggest the following:

   • Squeeze all water/air out of the dropper.
   • Lower the dropper into the water and "unsqueeze."
   • Lift the dropper, hold it over the slide, and gently squeeze.

3. Be sure students use the plastic depression slide with the flat side down. Pond water should be placed into the small "bowl" on the side facing up.

4. Students may have trouble with material getting stuck in the medicine droppers. If this happens, just pull up some water into the medicine dropper, gently tap the animals or other material into the water, and then squirt them out.

5. If there is too much water in the slide, have the students pull off a little with the medicine dropper so the pond organisms are confined to a smaller space. This will allow students to see them more easily and will also guard against spills.

6. For this station, a dissecting microscope is great. If you have only compound microscopes, be sure to use the lowest power. If some of the specimens are thick or opaque, have students move the light so it shines on top of the specimen, as well as up through the bottom of the microscope stage.

7. Water and microscopes are a bad combination. Please be sure to have students wipe up spills immediately. Make sure they wipe off the bottom of the plastic slides before they put them onto the microscope stage.

THE DISCOVERY QUILT

Overview

The Discovery Quilt activity is a summary station, which can be used during the learning station festival, or as a separate whole-class activity after the festival. If used during the learning station activities, this station can serve as a "sponge" activity for students who finish a station activity early. If the quilt is used as a whole-class, closing activity, it works well to have students share their reflections and questions orally before making the quilt.

Students create the Discovery Quilt by posting their findings, questions, and reactions to the learning stations. They are asked to write on colored, square pieces of paper what they have seen and learned during the festival, as well as what questions they would like to explore further. They tape each square to a large piece of butcher paper.

Through reflection on their experiences, students (and learners of all ages) are better able to construct new and improved understandings of what they have discovered in their investigations. The process of filling out a quilt square will help your students to reinforce and further apply the skills and concepts they have learned during the festival. Additional suggestions for reflection and closure are listed on page 102. "Assessment Suggestions" for the entire *Microscopic Explorations* unit are described on page 144.

One teacher commented, "The Discovery Quilt is a wonderful way for quieter students to communicate their learning."

If the Discovery Quilt option seems elementary for older students, choose another option for providing closure to the festival. Some ideas are listed on page 102.

What You Need

- ❏ a variety of markers, pens, and pencils
- ❏ 1 roll of masking tape (scotch tape is also fine)
- ❏ butcher paper or chart paper (at least 24" x 32")
- ❏ wide-tip marker, yardstick for drawing grid
- ❏ small construction paper squares in a variety of light colors (4" x 4" is a good size, but the exact size and shape are not critical)
- ❏ Discovery Quilt station sign

Getting Ready

1. Use a wide-tip marker and a yardstick to make a grid of 4" x 4" squares on the butcher paper (or to accommodate whatever size construction paper squares you have).

2. Gather together the small construction paper squares. If these have been cut carefully, they will fit well on the squares of the grid. (It's also possible to use paper of assorted shapes and sizes, and have students tape them up on a sheet of paper without fitting to a grid.)

3. Tape the butcher or chart paper to the wall.

4. Place the paper squares, markers, pens, or pencils, tape, and station sign on the table.

Discovery Quilt

What to Do

1. Take one of the squares of paper and write about something you discovered. Add a drawing if you wish.

   - What did you find out from doing these activities?

   - Did you see something that you have never seen before?

   - What was your favorite thing in the festival?

   - What would you still like to find out?

2. Tape your square to the Discovery Quilt.

3. If there's time, read some of the other student's discoveries and consider their questions.

Other Options for Reflection After the Festival

Use some of the ideas below—in combination with the Discovery Quilt or independently—to encourage students to reflect on what they learned at the Microscopic Explorations *Festival and consider future investigations.*

1. Have a discussion about something that surprised the students or that they had never seen before. Ask them to review what they recorded in their booklets, and then write in more detail about their favorite stations and what they liked about them.

2. Use the questions and observations they made on the Discovery Quilt as a starting point for a class discussion and continuing class, group, or individual investigations. You could also use the mural of pond organisms from the Pond Life station, or the map from the Sand station to spark their memories about what they observed.

3. Ask them why they think it might be important to see some of the things they saw with a microscope rather than just looking with their eyes. How would they compare the hand lens and the microscope? What other questions or investigations might require the use of a microscope?

4. Ask the students if they have further questions they would like to investigate. Ask how they might gain more information about these questions. If time and curricular emphases allow, consider providing students with opportunities to further refine and actually explore these questions.

5. Choose one of the many GEMS units listed in the "Going Further" sections for each station to deepen skills and concepts gained in the festival. In addition, the GEMS unit *Learning About Learning*, for upper elementary and middle school students, includes several main activities based on the brain research of Dr. Marian Diamond. Students simulate aspects of this research, including a counting process to measure dendritic growth. The drawings students analyze to do this are adapted from actual microscopic images derived from the research.

6. If you have volunteers who work with microscopes in their professions, have them share a short explanation of what they do in their work and how they use microscopes. Photographs or samples from their labs make their sharing much more interesting for the students. Leave time for students to ask questions.

7. Encourage students to do further research on the invention and refinement of the microscope and the contributions that microscopes have made and continue to make to new breakthroughs in human understanding. The book *Microbe Hunters*, listed in "Literature Connections," is a classic which documents the scientific discovery of microbes. A number of other books, for advanced students and adults, that describe scientific research in detail are also listed in the "Literature Connections" section.

8. Have students write a story on their adventures in a microscopic world. You may want to encourage students to exercise their creative imaginations, as well as provide detailed descriptions of at least three of the organisms or substances they observed, discuss how a microscope works, and give a sense of the relative scale.

9. Consider presenting one of the units or activity collections in the "Resources" listing. Based on student questions and your own curriculum plans, you may also want to consider several of the CD-ROM or videotape programs.

This GEMS guide was sponsored by a contribution from the Microscopy Society of America (MSA). On these two pages, we provide information about the MSA and ways for you to contact chapters in your region. The MSA is committed to helping educators obtain microscopes for use in Microscopic Explorations *and many other educational activities. MSA members can also be of great service as volunteers during the festival and/or as classroom visitors after the festival to talk with students about their work as scientists and how they use microscopes.*

THE MICROSCOPY SOCIETY OF AMERICA (MSA)

The MSA is an internationally recognized professional society and is the world's largest organization concerned with microscopy. In 1942, in the early days of the development of electron microscopy, a small group of scientists and engineers founded The Electron Microscopy Society of America (EMSA). Over the years, the Society has not only grown in size, but has broadened the "scope" to cover all areas of microscopy. In 1993, the name of the Society was changed to The Microscopy Society of America to better reflect and encompass the diversity of microscopies and associated techniques that have evolved in the last fifty years. In addition to instrumentation and techniques, the Society is concerned with a wide range of applications in both physical and biological sciences, as well as the development and use of all forms of microscopy, imaging, and compositional analysis. Currently there are over 4,000 members, including about 450 student members. The membership reflects a healthy balance among all aspects of the field. The Society is particularly interested in encouraging the activities of young scientists. The Society holds an Annual Meeting, often in conjunction with other societies, and carries out a number of activities and services for its members throughout the year. The Society represents the profession in public affairs, and serves as a center for education and information dissemination for microscopy-related knowledge. In addition, there are many meetings and activities organized by the 30 Local Affiliate Societies.

For further information, contact:
MSA Business Office
230 East Ohio St., Suite 400
Chicago, IL 60611-3265
Phone: (800) 538-3672 or (312) 644-1527
Fax: (312) 644-8557
Email: BusinessOffice@MSA.Microscopy. Com
WWW:http://www.MSA.Microscopy.com

MSA'S PRECOLLEGE EDUCATIONAL OUTREACH ACTIVITIES

Project MICRO (Microscopy In Curriculum–Research Outreach) is MSA's educational outreach program for middle schools. A collaboration between Project MICRO and LHS GEMS has produced this teacher's guide, and MICRO will help locate microscopist-volunteers to help you with classroom microscopy. Several pages on the MSA web site support *Microscopic Explorations*:

http://www.MSA.microscopy.com/ProjectMICRO A comprehensive bibliography of children's books, videos, CD-ROMs, and web sites. (Also available in print; request from the MSA office.)
http://www.MSA.microscopy.com/Ask-A-Microscopist.html This is an "ask-a-microscopist" help service for teachers.
http://www.MSA.microscopy.com/MicroScape/MicroScape.html An image quiz for children, and, in the near future, supplemental images for all of the *Microscopic Explorations* stations.
http://www.MSA.microscopy.com/MSALAS/LASInfo.html Current officer addresses for the MSA Local Affiliate Societies, your source of Project MICRO microscopist-volunteers.

MSA'S LOCAL AFFILIATE SOCIETIES

MSA is formally affiliated with 30 local and regional societies throughout the United States. Some of these societies have organized Project MICRO outreach program (see the MICRO website for a current list) and others will be able to locate microscopists who can help you present a *Microscopic Explorations* festival. **You can get current officer addresses from the MSA web page. If there is no local society near you, you can ask for microscopist-volunteers via the "ask-a-microscopist" service on the MSA web page; your request will be placed on the microscopy listserver, which is read by thousands of microscopists worldwide. Please try to make your request for volunteers well in advance of your need.**

ALABAMA IMAGING and MICROSCOPY SOCIETY
APPALACHIAN REGIONAL MICROSCOPY SOCIETY (east TN region)
ARIZONA IMAGING and MICROANALYSIS SOCIETY
CAPITAL DISTRICT MICROSCOPY & MICROANALYSIS SOCIETY (Albany, NY area)
CENTRAL STATES MICROSCOPY & MICROANALYSIS SOCIETY (southern IL area)
CHESAPEAKE SOCIETY of MICROSCOPY (DC and suburbs)
CONNECTICUT MICROSCOPY SOCIETY
DELTA MICROSCOPY SOCIETY (Delta College, Stockton, CA)
FLORIDA SOCIETY for MICROSCOPY
IOWA MICROSCOPY SOCIETY
MADISON AREA TECHNICAL COLLEGE SOCIETY
   for ELECTRON MICROSCOPY (Madison, WI)
METROPOLITAN MICROSCOPY SOCIETY (NYC suburbs)
MICHIGAN MICROSCOPY and MICROANALYTICAL SOCIETY
MICROSCOPY SOCIETY of NORTHEASTERN OHIO
MICROSCOPY SOCIETY of the OHIO RIVER VALLEY (southern OH)
MICROSCOPY SOCIETY of PUERTO RICO
MIDWEST MICROSCOPY and MICROANALYSIS SOCIETY (northern IL & neighboring states)
MINNESOTA MICROSCOPY SOCIETY
MOUNTAIN STATES for ELECTRON MICROSCOPY (CO, UT)
NEW ENGLAND SOCIETY for MICROSCOPY
NEW YORK MICROSCOPICAL SOCIETY (NYC area)
NEW YORK SOCIETY of EXPERIMENTAL MICROSCOPISTS (NYC)
NORTH CAROLINA SOCIETY for MICROSCOPY and MICROBEAM ANALYSIS
NORTHERN CALIFORNIA SOCIETY for MICROSCOPY
OKLAHOMA MICROSCOPY SOCIETY
PACIFIC NORTHWEST MICROSCOPY SOCIETY (OR, WA, ID)
PHILADELPHIA SOCIETY for MICROSCOPY
SOUTHEASTERN MICROSCOPY SOCIETY (Georgia & neighboring states)
SOUTHERN CALIFORNIA SOCIETY for MICROSCOPY and MICROANALYSIS
TEXAS SOCIETY for MICROSCOPY

This is a sixth grade class from the Essex Middle school, Essex, Vermont, who experienced "Project MICRO" as part of the New England Society for Microscopy's outreach efforts.

Behind the Scenes

Station Background

Up Close

The word "optics" comes from the Greek word *optikos*, meaning "of the eye or seeing." In science, optics is the study of light, and includes all the information human beings have collected about light and its behavior over many thousands of years of observation, experiment, and theory. One important portion of experimental optics has to do with lenses, how they affect the behavior of light, and how that in turn changes our visual (and mental) perceptions. Student experiences with lenses, both at this station, and through their use of hand lenses and microscopes at the other stations, can contribute a great deal to their eventual more theoretical understanding of the important physical science concepts related to optics. If you or an interested student or group of students would like to pursue this fascinating field, we provide a set of activities, beginning on page 123, in the "Special Section on Optics" by Professor Michael Isaacson of Cornell University, to enable you to investigate further. There are also many excellent resources for learning more about optics and microscopes in the "Resources" section on page 131. The GEMS guide, *More Than Magnifiers* is a great way to explore lenses and their properties, and includes good background information. Two other GEMS guides relate to light: *Color Analyzers* explores aspects of light and color, with background on the wave theory of light; in *Global Warming and the Greenhouse Effect*, a game relates to the particle theory of light.

Light travels in straight lines. But, if you put any clear material that is more dense than air (water, glass, plastic) at some angle to a light beam, the light will bend upon going into that material from the air. This is what a lens does. A lens is a curved piece of clear or transparent material that bends light; a magnifying (positive) lens curves outward (is convex). Lenses are usually made of glass or plastic. The light bends more or less depending upon the curvature (or surface angle) of the part of the lens it strikes. Light entering the center of a lens goes straight through, but light entering near the edge gets bent inwards towards the center. If the curve is right, there is a point to which all light coming from very far away gets bent or "focused." This focus is called the "focal point" of the lens and the distance of this point from the lens is called the "focal length." When lenses are stacked together, depending on their degree of curvature, magnification can be increased. A telescope is an instrument that magnifies distant objects. A microscope is an instrument that uses a combination of lenses to produce magnified images of small close objects.

How do magnifying lenses, or the lenses in a microscope, make things bigger? First of all, we only see things because light rays coming from those things land on the retina of the eye. The amount of space the image of an object takes up on the retina tells the brain how to interpret the size of the object. If an image takes up a lot of space on the retina, the brain assumes it is larger. So, getting back to a magnifying lens, when it bends the light, it causes the light rays to occupy a larger space on the retina.

"Where the telescope ends, the microscope begins. Which of the two has the grander view?"

Victor Hugo,
Book 3 Chapter 3
Les Miserables

"Nature composes some of her loveliest music for the microscope and telescope."

Theodore Roszak
Where the Wasteland Ends,
1972

That is why people say lenses "fool" the eye—the object of course retains its actual size, but because of the way the lens bends light, we perceive of the object as larger—and can actually see it in much greater detail! (There are also reducing lenses, with concave surfaces, that make an object appear smaller.)

By bending light before it gets to our eyes, lenses change what we see. When light rays from an object pass through a magnifying lens, they cross. After they cross, the light rays that were coming from the top of the object are now on the bottom, and the light rays coming from the bottom of the object are on top. Similarly, the left and right sides of the object are reversed. The eye registers this reversed image and we see the object upside-down and backwards (called an "inverted" image). If you start with the object very close to the lens, however, the light rays do not cross and the image is not inverted. As you move the lens outward, the point at which the image "flips" is the focal point.

The eye itself also contains a lens. The curved clear surface at the front of the eye and the lens inside the eye both bend light to make an image that— if we are neither near-sighted or far-sighted—focuses on the retina. This image is then transmitted through the optic nerve to the brain. The image that is focused on the retina is also upside down when on the retina, but our brain proceeds to interpret it as right-side-up. We learn to do this image "flip" in the first few days of life. The eye, optic nerve, and brain also interpret the image in terms of color and depth, and we see the marvels of the world around us. If we use a hand lens or a microscope, thanks to the way lenses bend light, we are able to see a vast world that we are unable to see with the naked eye!

Fingerprint Ridges

Because no two fingerprints are identical, matching a fingerprint found at a crime scene to a suspect's print can be powerful evidence in a criminal case. Crime lab scientists are often called upon to testify in court about whether two very similar fingerprints are—or are not—identical.

To determine whether two prints really match, experts may need to go beyond using the standard fingerprint patterns (such as arch, loop, whorl). In order to feel confident enough to testify about whether two very similar prints are identical, the experts may need to compare individual, tiny characteristics of the lines or "ridges." They look for certain matching ridge patterns, such as forks, islands, and other characteristic details. (The seven ridge details on the student booklet and station sign are the most common; more are used by experts.) When the crime scene print is partial or smudged, studying the ridges can be especially valuable. Some experts feel comfortable testifying in court that two prints match when they have identified eight or nine matching ridge details, and others need more. In England, for example, 23 matching ridge characteristics are required to determine that two fingerprints are identical.

To focus on the print ridges, crime lab scientists often use magnifying lenses (3x to 5x magnification is usually most helpful), aiding their eyes by tracing the ridges with a fine-tipped pointer. Experts work in teams, always confirming their observations with a colleague. In many cases, they photograph two fingerprints they are comparing, enlarge them, and project them on a special screen. Sometimes experts create poster-sized exhibits of the enlarged prints with key ridge characteristics labeled. Such exhibits can be used in court to help juries determine whether two prints are identical or not. Here are some references for more information:

The Crime Laboratory: Case Studies of Scientific Criminal Investigation by James W. Osterburg, Indiana University Press, Bloomington, Indiana, and London, 1967.

The Science of Fingerprints by U.S. Department of Justice, FBI, available from the U.S. Government Printing Office, or from Lighting Powder Company, 1230 Hoyt Street SE, Salem, OR 97302-2121. (800)852-0300.

The Boy Scouts Handbook, The Boy Scouts of America, Irving, Texas, 1990.

Dots and Dollars: Color Printing and Half-Tones

At the Dots and Dollars station, students are exploring part of the technology of color printing, in which all colors can be created by the mix of four main colors. The four "process colors" used in printing are: cyan (a greenish blue) which absorbs the red part of white light; magenta (a purplish red) which absorbs the green part of white light; yellow (a pure yellow) which absorbs the blue part of white light; and black, which absorbs all parts of white light. In color theory, the first three of these colors (cyan, magenta, yellow) can be blended in varying combinations and dot sizes to make all the colors of the spectrum, as well as black. However, printer's inks are not and cannot ever be perfectly matched to all the theoretical hues, so a fourth "color," black, is added to help produce the darker shades of color and the pure black of shadows. When the three main process colors are printed *over the same area with dots of the same size*, the result will be a gray color. But if any one of these is altered, there will be a color change. For example, a yellow ink (which reflects red and green) when overprinted with an equal amount of cyan (which reflects green and blue) will give the appearance of green. In this case, the yellow does not permit the blue to be reflected, and the cyan does not permit the red to be reflected, so only green can be seen.

Four-color process printing gives an illusion of continuous color tone in natural colors. However, as your students discover, if you look at a color illustration in a magazine under a magnifying lens, you see that the image is actually made up of tiny dots of these four colors, some overlying each other, and many of them adjacent to each other. The dots of each color vary in size and number across the image, and our unaided eye, which cannot distinguish the individual dots, interprets the patterns made as natural colors. During the production process, through the use of appropriately colored filters, negatives of the dots of each of the four colors are made—this is called a color separation process. Printing plates are then made from these negatives. Pages can be printed by making four passes through a single-color press, or, as is much more common and economical, by one pass through a four-color press. The front and back cover of this guide was printed in this way. With the advent of the computer and color printing, the technology of color reproduction is changing rapidly. The four-color process for scanning and printing has become more computerized, but is still based on these combinations of tiny dots. **The book by Ruth Heller entitled *Color*, listed in "Literature Connections," does a superb job of demonstrating this process. We highly recommend it.**

Students may observe that reprinted photographs and some other non-color graphics are also made up of tiny dots. Printing plates (unlike photograph paper) can record only pure black and its absence (pure white) and are not able to record intermediate shades of gray. In order to re-create the continuous tone and appearance of a photograph, which has many such intermediate shadings, a screening process is used that represents the photographic image as a half-tone with tiny dots. The size of the black dots depends on the tone of the photographic area. The black dots in the lightest areas are very small; they are somewhat larger in medium-gray areas; and in the darker areas they are so large that they merge together. As in the four color process, the dots are too small for us to see with our eyes alone, and we perceive an illusion of black and white and shades of gray blending together in a continuous tone, as in the original photograph.

As your students will discover, the one dollar bill is not printed with tiny dots, but instead has black and green lines and solid areas. George Washington's face is made with dashes in rows; his hair of black lines; his jacket of black lines criss-crossing each other. There are a few tiny blue and red "hairs" on the front and back of the one dollar bill—these are added to try to make it more difficult to counterfeit them. Other U.S. currency also has some interesting attributes. Our currency has tiny microscopic words in several places; finding them all is a popular challenge (see photograph on front cover).

Some postage stamps are printed using color dots, some use tiny lines, and others display other methods of printing and design. Invite your students to see how many different methods they can find.

Fabrics

The oldest form of fabric is wool felt, which is made by pressing wool fibers together. Sheep's wool has scales on the fibers that help the felt stay together. The process of pressing includes using soap and moisture and heat to help create a permanent bond. Interestingly, felt was probably discovered when people put wool in their shoes under their feet to keep warm, and voila!—fabric.

Historically, woven fabrics came next. Threads are spun, often with a mixture of wool and animal fibers (like rabbit, goat, alpaca, even raccoon) or plant fibers (cotton, hemp, wisteria, banana, or whatever grows in the region). One set of threads is fixed taut on a frame (the warp threads). The weft threads are then woven perpendicular to the warp. This was done by hand at first, but people quickly figured out how to mechanically lift alternate warp threads (with something called a "shed") so that a weft thread can be quickly passed through. A "fly" shuttle sends the weft thread between the warp threads. There are lots of ways to weave the weft; it can be looped around each warp thread or can skip certain numbers of warp threads to form different patterns. Warp threads are usually made with stronger fibers than weft threads because they need to withstand stretching during the weaving process. Modern industrialization has of course created many new and rapid ways to interweave fabrics and the development and technologies of these industries would make an excellent research topic.

Although ancient peoples devised many techniques of knotting and netting, modern knitting originated after the development of woven fabrics. Unlike woven fabrics, knitted fabric is made with one continuous thread. Each row of loops interlocks with the row before it. There are many ways the loops can be varied to create patterns. Of course, fabrics have long been knitted by hand using knitting needles, but knitting machines now create the miles of fabric needed to clothe the modern populace in T-shirts, sweaters, socks, and other knits. One excellent resource book on textiles is *Women's Work: The first 20,000 years; women, cloth, and society in early times* by Elizabeth W. Barber, Norton, New York, 1994.

Salts: Crystals and Crystal Formation

Crystals are familiar objects, prized for their shapes, for their beautiful, multi-faceted reflections of light, and thought by some to possess spiritual and healing qualities. Distinctive crystalline shapes are a defining feature of many minerals. Scientifically speaking, crystals are defined by their structure. To be precise, a crystal is an homogenous solid with a regular geometric form, whose boundaries are naturally-formed smooth, planar surfaces, also called faces. This regular geometric form reflects an orderly three-dimensional internal structure. Mineral crystals can be formed from solutions, from melts, and sometimes from vapors, due to changes in temperature, pressure, or concentration. When this happens, atoms that are in a more random or disassociated state are combined together in a more orderly arrangement. Salt water is a good example of crystal formation from a solution. As the solution evaporates, the concentration of sodium and chloride atoms increases to such an extent that these elements group themselves together. The halite crystals and sodium chloride molecules precipitate and salt crystals are formed. Crystals are also formed through melting. Igneous rocks, formed from molten magma, begin to cool. As cooling continues, atoms arrange themselves in a definite order, forming nuclei of different minerals that eventually form a solid mass of crystals. Crystals can also be formed from a vapor, although this is less common. As the vapor cools, disassociated atoms are brought closer together, eventually locking themselves into a crystalline solid.

There is a fascinating mathematical "facet" to crystalline shape—there are only a limited number of crystal shapes possible. These have been classified into seven systems, or classes of symmetry, with 32 sub-groupings.

The GEMS guide *Stories in Stone* features much more information on crystals and crystal formation, including activities that: model crystal formation in igneous rocks; explore the creation of salt crystals through evaporation; and re-create different crystal shapes (including halite, or salt) with paper models. *Stories in Stone* also includes detailed background information and several poems of Chilean poet Pablo Neruda, including one in which he refers to diverse crystalline shapes as a "whole buried geometry." In *Microscopic Explorations*, your students gain direct experience with several kinds of salt crystals at the Salts station, and learn that different salts have distinct crystalline shapes. *Stories in Stone* would make an excellent in-depth follow-up unit for the Salts station.

Sand

Geologists classify the sediments in the earth by particle size. Grains of clay tend to be flat, and are less than .002 mm (1/12,500th of an inch) in diameter. Silt is up to 0.5 mm (1/500th of an inch) in diameter, and sand ranges from .1–2 mm (1/250th of an inch to 1/12 of an inch) in diameter. Larger pieces are often called pebbles. Materials with grains larger than sand are often referred to as "coarse," while those smaller than silt are referred to as "fine."

Nearly all the solid materials in the world, living and non-living, will eventually be eroded into sand. Mountains, rocks, minerals, shells, coral, bones, metals, and glass are all worn down over time by wind, waves, rivers, earthquakes, and other forces into smaller and smaller particles. The story of a grain of sand can reflect the story of the evolution of the Earth's crust.

The sand on every beach has its own unique history. Detailed observations can provide much information on the possible origins of the sand, how old it is, what geographical location and part of the beach it was collected from, whether it is biogenic (includes remains of plants and animals) or not, as well as highlighting the uniqueness and beauty of individual grains. Some sand is produced at the shore, as waves crash on rocks, headlands, and reefs, such as the black or red sand beaches on the Hawaiian and Galapagos Islands. White sand beaches in Florida and the Caribbean are primarily made up of eroded coral reefs. Other sand comes from far inland, as mountains are weathered by running water, ice, wind, and rain and fragments come down streams and rivers to the shore. Quartz is the most common component of these transported sands, and most light-colored beaches contain large amounts of quartz. Where particles are deposited depends on their size and the speed of the water carrying them. If sand grains are very small, for example, they may be from an area with slow moving water, such as a protected bay beach or pool in a slowly moving stream, because such small particles can stay put only where the water moves slowly and gently. If the sand grains are large, they may be from a wave-tossed beach where the rough water carried all the smaller grains away.

On an average coastal beach, no individual sand grain stays in the same place for long. Each wave picks up many thousands of grains and deposits them elsewhere. The sand on a beach is in constant motion—sand grains may be hit by as many as 8000 waves a day! Because waves hit the beach at an angle, the sand is also moved along the beach, often for considerable distances. The finest grains of sand can become airborne in the wind, and are often deposited high up on the beach, so dune sand is finer and lighter than beach sand. Sandy shores are one of the most unstable marine environments—much more accurately described as constantly moving rivers of sand, rather than anything permanent. The GEMS guides *On Sandy Shores*, *Stories in Stone*, and *Terrarium Habitats* all have activities and background that focus on sand and/or the composition of soil.

Brine Shrimp

Brine shrimp can be found in nearly all parts of the world and occur widely across North America. They live wherever there is very salty or saline water, including coastal salt ponds and inland salt water lakes—such as the Great Salt Lake in Utah and Mono Lake in California. As zooplankton, they occupy an important place near the base of the food pyramid—eating phytoplankton (single-celled algae and other minuscule plants) and being eaten by very young fish, filter-feeding invertebrates, and many shorebirds. Captive brine shrimp can be fed a very small amount of bran or powdered, dried yeast once or twice a week.

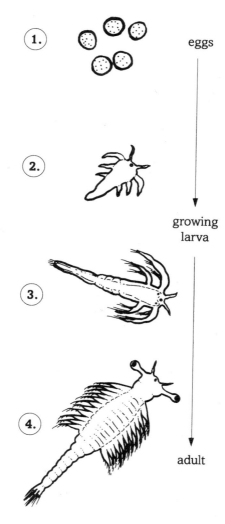

1. eggs

2. growing larva

3.

4. adult

Brine shrimp are crustaceans, like crabs and lobsters. All of a brine shrimp's hard parts (the exoskeleton) are on the outside. This hard shell doesn't stretch as the animal grows, so the brine shrimp must shed the old shell (molt) when it gets too tight. The new shell growing underneath is still soft when the old one is shed, and so it expands, giving the brine shrimp room to grow. Brine shrimp have a mouth, two compound eyes (like an insect's) on stalks, and a third small eye in the middle of the head. The gills are located on each of its 22 legs (11 pairs). To supply itself with oxygen, a brine shrimp must keep water moving over its gills. Very salty water has less oxygen than fresh water so a brine shrimp must find a way to store oxygen in its body. To do this, it produces hemoglobin, just as in human blood, which makes it look red/orange. Brine shrimp can become different colors depending on the color of the food they eat!

The "horns" on the male are used during mating to hold the female while the male fertilizes the eggs in her egg sack. Young brine shrimp may hatch inside the mother's egg sack and be released live, but more often the female releases the eggs before they hatch. Each egg is encased in a cyst—a thick-walled sphere that contains and protects a dormant embryo. The cysts tend to float and drift to the shore, and must dry before hatching. They are able to withstand long dry spells until conditions are suitable for growth. When the egg hatches, the free-swimming larvae have no tail or legs; these grow in later. As the larvae grows, it sheds or molts its exoskeleton seven times before reaching its adult size and shape—all in about six weeks. Adult brine shrimp can live for two to three months and eggs can develop when the female is three weeks old. The illustration depicts stages in the brine shrimp's life cycle.

Pond Life

We've all heard the phrase—"that pond is teeming with life." We readily accept this statement because we can see many life forms in and around the pond—birds, plants, insects, and fish. But the phrase "teeming with life" becomes even more apt when just a single drop of pond water is viewed through a microscope. In that drop, an amazing variety of animals and plants—ranging from single-celled to fairly complex organisms—can be found. Examples include bacteria, protozoa, algae, diatoms, amoebas, rotifers, and arthropods.

In the Pond Life activity, students are likely to see two common organisms—a plant called *Elodea* and an animal called *Daphnia*. *Elodea*, also called Anacharis or water weed, grows entirely submerged as a loosely rooted or free-floating plant. The branched stems are crowded with green, translucent, narrow leaves arranged in whorls of three or more. *Elodea* spreads with amazing speed and may literally fill up a pond or slow stream and crowd out other plants.

Daphnia, also often referred to as water fleas, are among the smallest members of the crustacean group. They have a shell which covers all of the body except the head and antennae. The shell is clear and all the internal organs can be easily seen including the prominent compound eye behind which is the brain. *Daphnia* have no blood vessels; the action of the heart moves blood around the body cavity. The intestine is seen as a dark tube running through the length of the body.

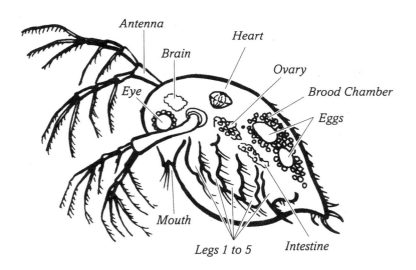

The primary food for *Daphnia* is algae. *Daphnia* have a groove or canal that runs along their stomach area, with five legs on either side that act as both fans and filters. The legs wave around in many directions making currents that move pieces of food along the groove. The legs also pack the food into a tiny ball. The last pair of legs pushes the ball of food into the mouth. As filter feeders, *Daphnia* play an important role in clearing water clouded with algae and protozoa.

Near the top of the head, *Daphnia* have antennae which look like branching arms. These antennae are used for swimming. By jerking the antennae, the *Daphnia* swims up through the water, then rests. When still, it begins to sink. Another jerk of the antennae moves it forward again. From their body shape and the jump-like way they move through water, *Daphnia* have earned the nickname water fleas. They swim freely about in open or shallow water among the plants at the edge of a pond.

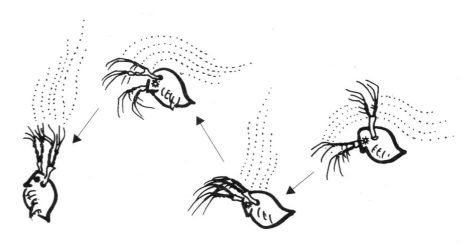

Most *Daphnia* are female and reproduce by parthenogenesis (the development of an egg without the benefit of male fertilization). When conditions are favorable, females can produce many eggs to increase the population and take advantage of a large amount of algae. But when environmental conditions are different, males develop and the mode of reproduction changes. The eggs produced then have a harder shell which can resist drought and live through the winter.

A good source of information on the large variety of microscopic animals found in ponds is *Pond Water Zoo: An Introduction to Microscopic Life* by Peter Loewer, Simon & Schuster, New York, 1996. Another useful book is *Guide to Microlife* by Kenneth Rainis and Bruce Russell, Franklin Watts, Danbury, Connecticut, 1996.

Compound and Dissecting Microscopes

There are two types of microscopes that we recommend using for *Microscopic Explorations*. Both have two sets of lenses in the long objective tubes, increasing the magnification. **Compound microscopes** are the traditional microscopes with which most people are somewhat familiar. Usually there are two or three different objectives, of different lengths and different magnifications. Sometimes there are interchangeable eyepieces of different magnification. For the festival, choose lower power. If you look on the side of the objective, it will have a number, followed by an "x," which gives the magnification. Common objectives have 1.5x, 10x, or 40x magnifications. Light comes from below the stage, and shines through the object. For this reason, samples need to be very thin or transparent to be seen well. Compound microscopes may have one or two eyepieces. Because of the combination of lenses, images appear upside-down and backwards, which can make chasing a small *Daphnia* quite a challenge until you are used to it!

The other type of microscope, the **dissecting microscope**, is designed for exactly what its name suggests. The purpose of these microscopes was originally to allow a scientist who was dissecting something to be able to see better. For that reason, an extra lens, mirror, or prism is added to show things in the correct orientation, so things appear right-side-up. These microscopes usually have two eyepieces, and the magnification is lower than that of compound microscopes. Also, because the light can be arranged to come from above, dissecting microscopes are useful for seeing non-transparent objects (fabrics, brine shrimp eggs, salts, kitchen powders, bugs, sand, and the objects at the Dots and Dollars station).

Both types of microscopes are useful, and both give different information. At several stations, we recommend that—ideally—both types be used, in order to allow students to learn as much as possible from the materials. If you do not have access to dissecting microscopes, you can change the lighting on the compound microscope to observe surfaces that are not transparent. Take a portable light source and arrange it so the light is shining from above. Use the lowest objective, and students should be able to see the surfaces well enough to describe them.

For students taking part in *Microscopic Explorations*, microscopes with two eyepieces (binocular) are not necessarily better. Sometimes the distance from eye-to-eye is too large for young faces, or there is difficulty fusing the two images into one. Monocular microscopes are less expensive, don't go out of alignment, and are generally easier to use. Please see the next section, "Selecting School Microscopes" for more information.

Selecting School Microscopes

There are a lot of inexpensive microscopes available, of widely varying quality. Some of the good ones cost no more than some of the nearly useless "toy" models. If you follow the basic evaluation criteria given here, you don't have to be an optical expert to make a good choice. Remember that "experts" don't always agree, and trust your own evaluation skills.

What kind of microscope should I buy?

The first choice is between "simple" and "compound" microscopes. A "simple" microscope (Leeuwenhoek used one) has just one lens; a "compound" scope has both an objective and an eyepiece. **Don't buy a "simple" design!** The working distances between eye and lens and lens and specimen are so small that they are very difficult to use. In addition, a single powerful lens has so much aberration that the student who manages to get an image will be disappointed by its quality. Unfortunately, there are quite a few models offered in school supply catalogs. You'll do better with a set of hand lenses or loupes (as used, for example, in the *Private Eye* materials, listed in the "Resources" section on page 132).

But I only have $100 to spend—does that mean no microscopy?

Not at all! Many manufacturers offer a design that looks like a pocket flashlight. They usually magnify 30x, with two AA batteries providing illumination, and they sell for from five to ten dollars. You'll find them in electronics and "nature" stores, and many catalogs. Their quality varies, so it's wise to compare. They can be good enough to support extensive curriculum (see for example *Experiments with the Mini-Microscope*, listed in the "Resources" section on page 132). Buy as many as you can afford; some local dealers may be willing to discount a bulk purchase for school use.

I want to equip my classroom with "real" microscopes—what will that cost?

Approximately $1000 (don't despair; see below). That will get you at least 10 good quality scopes in the $50–$100 price range. You can get scopes at that cost that will be durable and easy to use, with lenses that will deliver a sharp, bright image. In general, more expensive models will provide similar images but more convenience, and less expensive ones will have disappointing performance.

What type should I buy?

Two types, actually, in roughly equal numbers. Inspection/dissection scopes are used to look at large, opaque specimens at relatively low (20x–30x) power. Illumination is usually from above, and *the image is "right-side-up,"* as in the "real world." Compound microscopes are usually used with transmitted light to look at transparent specimens; the useful school magnification range is 10x–400x. *The image is inverted.* It takes a bit of

"...can the human soul be glimpsed through a microscope? Maybe, but you'd definitely need one of those very good ones with two eyepieces."

Woody Allen

"There is no advantage in examining any object with a greater magnifier than what shows the same distinctly..."

Henry Baker,
The Microscope
Made Easy,
1742

practice to follow a moving subject when it's upside-down; the CD-ROM *Scopemaster* provides helpful practice (see the "Resources" section, page 141).

What features should I look for?

Both types should have metal bodies and metal rack-and-pinion focus, for durability and easy, precise focusing. That immediately eliminates the plastic "toy" scopes. Although a metal body is no guarantee of lens quality, metal focus gearing is more precise than twistable plastic designs. Both types should have glass rather than plastic lenses and be able to focus on both thin specimens (slides) and larger objects, at least an inch thick. Compound scopes should have a three-lens turret and a substage diaphragm (light control) or series of "field stops" to control brightness. There are some good single-objective compound scopes available, but the three lens design is much more versatile; a student can locate a specimen at low power and immediately switch to higher magnifications.

I see a lot of other features advertised; which ones are worthwhile?

"Magnifies 600–1200 times!" NO. When you see this claim in an advertisement, it's good reason to read no further. The wavelength(s) of visible light and the optical properties of glass lenses used in air (rather than the "immersion oil" used with research microscopes) limit the useful magnification of a compound school microscope to 400x; more is "empty magnification." Higher magnifications are achieved in "toy" microscopes by using an eyepiece of excessive power, which in turn makes the field of view very narrow, while emphasizing all the aberrations of the image produced by the objective lens. It's like enlarging a snapshot from a cheap camera to poster size; it's bigger, but there's no more detail. **Most school microscopy needs 10x–100x** (bacteria require 400x).

"Zoom magnification!" NO. This is related to the preceding problem. A zoom eyepiece just makes things worse, because cheap zoom optics are full of aberrations. Magnification changes in a compound microscope should be accomplished by changing objective lenses, not by zooming the eyepiece. That change is best accomplished with a lens turret rather than changeable screw-in lenses, which are easy to damage or lose.

Binocular eyepieces. NO. Two sets of optics cost a lot more, and if rough use knocks them out of line, factory service is needed. In addition, children's narrower interocular distance often won't fit adult eyepiece spacing. And, statistically speaking, fully 17% of children (5 or 6 in a class of 30) are likely to have amblyopia, strabismus, or other binocular coordination problems.

Condenser. NO. Although the substage condenser lens, which focuses illumination on the specimen, is an essential part of a research microscope, it should be avoided in the $100 price range. If one is offered it will be a single lens which can't be focused, fixed in the stage. It will be easy to damage and difficult to clean.

"Projection microscopy!" MAYBE. Even if you can completely darken your room, illumination sufficient to project an image with one of the cheap direct projection scopes will also "fry" your specimen. One or two manufacturers do have good, educationally useful video projection systems, but their cost makes it doubtful that they're "worthwhile" if the budget is limited. If you want to show an image to a whole class, mount a video camera an inch or two above the eyepiece. Use a camera tripod or other separate stand to hold the camera, with a piece of black foam pipe insulation (it's vibration-free, adjustable, and cheap) as a connector. Try both close-focus and infinity camera focus settings.

Built-in illumination. MAYBE. There's a lot of disagreement on this one. It's always more expensive than a mirror, and if every scope has an electric cord, the floor will look like spilled spaghetti. If you need illumination from above ("incident" lighting) for an opaque sample, a below-stage lamp won't provide it. Many built-ins require hard-to-get bulbs. Built-in battery-powered illuminators are offered on a few scopes, but have short battery lives. If scopes have mirrors, you can use any kind of table lamp. A short fluorescent tube in the center of the table can illuminate several scopes; they're about $10 at hardware stores. On the other hand, some children may have difficulty setting a mirror properly. If you do opt for built-in illumination, remember that wet samples can be a shock hazard and equip your electric outlets with ground fault interrupters (GFIs), which automatically cut the power to a "short circuit." Recently built classrooms may already have built-in GFIs.

A new type of illuminator, the light-emitting diode (LED) is now available for school microscopes, and it's definitely worth considering. LEDs cost no more than conventional lamp illuminators, the emitter itself lasts thousands of hours, and they're powered by rechargeable batteries that last over 50 hours between charges; the illumination is bright and even.

"Made in the U.S.A." MAYBE. If this is important to you, check directly with the manufacturer, not the dealer. Many American-brand microscopes are imported, and even scopes that are advertised as made in the United States may have important imported components, such as lenses.

Widefield eyepieces. YES. These provide a large, bright image and are usually the best choice. They let you see more specimen area than a conventional eyepiece of the same power. This also means that more illumination is gathered and transmitted, providing a brighter image. They should be no more than 15x; 10x is preferred. In student scopes, they're often fixed in place, which protects against loss, damage, and internal dirt.

Fine focus. YES. Desirable, but scarce in the low price range.

Focus stop. YES. This will prevent slide or lens breakage. If the scope doesn't have fine focus this is particularly important. If you can't focus on a very thin specimen (paper, or a plastic slide) you may have to put the specimen on top of a glass slide.

Can I check lens quality?

You can tell a lot without test equipment. The rectangular engraved crosshatching around the portrait heads on U.S. currency (best on the one dollar bill as shown on front cover) is a useful specimen for a crude check of lens quality. Focus up and down and look for several things:

A flat focal plane. You can't expect inexpensive lenses to have a perfectly flat field; so-called "plan" lenses are costly. You will, however, find a lot of variation in quality; try to do a side-by-side comparison if you're choosing between two models. You're looking for an image that's really sharp from the center almost to the edge of the field of view, rather than one that must be refocused for each part of the circular field.

Achromatic lenses are highly desirable. This means that the lenses will focus one wavelength well (usually mid-green, where our eyes are most sensitive). Uncorrected lenses will show color fringes around specimen detail. Apochromatic lenses, which are corrected for three wavelengths, are too expensive to consider here.

No distortion. The horizontal and vertical engraved lines should be straight.

No astigmatism. As you go through the focal point (fuzzy-sharp-fuzzy), look for a "linear" image distortion that rotates 90 degrees as you go from above focus to below focus. This can be caused by an objective lens that isn't round or (more common in cheap optics) a lens that is tilted in its mount. Rotate the eyepiece to check for the same problem with that lens.

No internal dirt. Defocus the image and look carefully for lens dirt on both objective and eyepiece. Try removing it with gentle use of lens tissue and small amounts of alcohol or eyeglass lens cleaner (Sparkle brand window cleaner works well). If it's *within* a multi-element lens, don't buy the scope.

Compound scope three-lens turrets should be reasonably well aligned and the cheaper ones often are not; compare the scopes that you're considering in this regard. Focus on a small centerfield object at the lowest magnification, and then rotate to the next higher power. The chosen object should be close enough to center at the new magnification so you can still see it, and it shouldn't be completely out-of-focus. Repeat for each objective, in sequence. Don't expect perfection; approaching that is expensive and will only be found in research scopes.

Try to do a scope comparison with two prepared slides. One should be a brightly-stained biological specimen and the other something that's almost colorless. Use different magnifications and illumination (the "field stops" mentioned above). The better lenses should make themselves apparent.

Where do I find these microscopes?

Major scientific supply catalogs and some of the school supply houses carry them. Check page 120 and the ProjectMicro web page (http://www.msa.microscopy.com/ProjectMicro/SourcesMicroscopes.html) for specific suggestions.

Where do I find the $1000?

That may not be as difficult as you think. Local corporations are often a good source of funding at this level. The MSA is prepared to help you write a grant proposal, and the MSA Local Affiliate Society nearest you may be able to help. (Please see page 103 for MSA contact information.)

What about buying a used microscope? I see a lot of great deals on EBay.

If you know enough to evaluate a used scope or repair a faulty one you probably won't be reading this basic advice. Microscopes don't "wear out"; they're often on the market because they aren't working as well as they should. Even if they're in good condition, the multiple controls on a used "research" scope will confuse and discourage beginners; school microscopes need to be simple to use.

Sources for Materials

Microscopes and other materials

Please see the section on "Selecting School Microscopes," beginning on page 115. The factors discussed there should help you in assessing the best microscopes for your school or district. Not all of the microscopes offered by the suppliers listed below are appropriate for middle schools. Don't assume that "more expensive" is always better. Look carefully at the abilities and experience of your users—and at their needs. You want to meet those needs, but you don't want to intimidate them with equipment that is too complicated. At the other end of the scale, you don't want to purchase low-quality equipment that won't do the job and will disappoint its users.

See also the ProjectMicro web site (http://www.msa.microscopy.com/ProjectMicro/SourcesMicroscopes.html) for a list of suppliers.

Beckley Cardy Group
1 East First St.
Duluth, MN 55802
(800) 446-1477

Carolina Biological Supply Company
2700 York Rd.
Burlington, NC 27215-3398
(800) 334-5551

Connecticut Valley Biological Company, Inc.
P.O. Box 326
82 Valley Rd.
Southampton, MA 01073
(800) 628-7748

Edmund Scientific Company
101 East Gloucester Pike
Barrington, NJ 08007-1380
(609) 573-6270

Fisher Science Education
485 South Frontage Rd.
Burr Ridge, IL 60521
(800) 955-1177

Frey Scientific
100 Paragon Pkwy.
P.O. Box 8101
Mansfield, OH 44901-8101
(800) 225-FREY

Insights Visual Productions, Inc.
P.O. Box 230644
Encinitas, CA 92024
(619) 942-0528 *or*
(800) 942-0528
www.microscopeworld.com

Nasco
901 Janesville Ave.
Fort Atkinson, WI 53538-0901
or
4825 Stoddard Rd.
Modesto, CA 95356-9318
(800) 558-9595

Northwest Laboratory Supply, Inc.
5510 Neilsen Rd., Unit B
Ferndale, WA 98248
(800) 469-4180

Science Kit & Boreal Laboratories
777 East Park Dr.
Tonawanda, NY 14150-6784
(800) 828-7777

Southern Precision Instruments
3419 E. Commerce St., Suite E
San Antonio, TX 78220
(800) 417-5055

Swift Instruments, Inc.
P.O. Box 562
San Jose, CA 95106
(800) 523-4544

VWR/Science Education
911 Commerce Ct.
Buffalo Grove, IL 60089-2375
(800) SARGENT

WARD'S Natural Science Establishment, Inc.
5100 West Henrietta Rd.
P.O. Box 92912
Rochester, NY 14692-9012
(800) 962-2660

Optics

The following materials are helpful for the optics activity for teachers in the "Special Section on Optics."

- Unmounted optical lenses/Double convex lenses
- Lens and mirror supports
- Optical bench set

These can all be ordered from the Nasco catalog (see address on previous page).

Clear plastic specimen boxes

Among other sizes, a 2" x 2" x ¾" high clear plastic box with a tight-fitting lid is available from

Althor Products
P.O. Box 640
Bethel, CT 06801
(800) 688-2693

Deli lids

If you are unable to obtain donated deli container lids from parents or a local deli or grocery store, you may order a large quantity of them from

AC Paper
1321 7th St.
Berkeley, CA 94710
Attn: Ted Robinson
(510) 527-0841

Product code: 736-40040
LG8 Solo Deli Lid
Case of 1000 (minimum order)

If you know of other large suppliers, please let us know.

Sand

You can obtain sand from a variety of sources. Try posting requests for sand throughout the school. You could request sand from other teachers, from your own family and friends, or from the family and friends of students. Let them all know that you don't need a lot of sand. You could, of course, collect from a nearby shore (lake, river, or ocean). Or you may want to try local sand and gravel companies, garden centers, hardware stores, or aquarium stores.

MSA's Project MICRO will supply sand to teachers; see the "Classroom Activities" section of the MICRO website.

Two different kits containing sand are available from:

Creative Dimensions
P.O. Box 1393
Bellingham, WA 98227
(360) 733-5024

The first kit is called Jewels in the Sand. It includes 10 samples of natural sand, each rich in a gem mineral. With each sand sample is a guide sheet, background information, and activity instructions. The second kit is called the Sand Activity Kit. It includes an illustrated teacher's guide, sand samples from seven different beaches, four sand samples from beaches on a spit, and 100 paper cups for sand activities.

Another option for obtaining sand may be the International Sand Collectors Society (ISCS). Their Fact Sheet notes that the organization was founded in 1969 by William S. Diefenbach to promote the collection and classification of sand samples excavated from world sites, as relates to hobbies and scientific interest. Though many society members exchange sand samples, the society itself does not arrange or conduct such exchanges, nor does it buy or sell samples or collections, but it does publish the surpluses as well as the wants and needs of its members. The society encourages the hosting of area meets so members can exchange ideas, information, and samples. These meets can be organized by any ISCS member and are the sole responsibility of that member. For more information, contact:

International Sand Collectors Society
P.O. Box 117
North Haven, CT 06473-0117
Nicholas F. D'Errico, Director/Publisher

Brine Shrimp

If you are unable to get brine shrimp from your local aquarium store, both eggs and live brine shrimp can be ordered from biological supply houses (such as Carolina Biological; address above). It is easier to ship eggs, but they must be hatched before they're used. Hatching the eggs requires specific conditions (see the "Two to three days before the festival" portion of the "Getting Ready" for the entire festival on page 22). A starter kit can be helpful for hatching the eggs. The biological supply houses usually ship certain organisms only on certain days. Call to find out on what day live brine shrimp are shipped and ask how soon you will receive them. Then plan to do Activity 9 as soon as the live brine shrimp are received because they don't live very long.

Pond Organisms

Please see the discussion concerning pond organisms in the "Getting Ready" section for the entire unit, under "One month before the festival." There you will find information on collecting and preparing pond water with organisms from nearby ponds or in water you collect. If you do not have such a source or are unable to prepare this yourself, pond organisms, such as *Daphnia*, can be ordered from scientific supply companies, including some of those listed at the beginning of this section under "Microscopes and other materials." Some aquarium stores may carry *Tubifex* worms, as well as plants, such as *Elodea*, and mosquito larvae may be available from a mosquito abatement agency.

Seurat Prints

Postcards of the Seurat painting "A Sunday on La Grande Jatte," (see optional materials in the Dots to Dollars station) are available for 65 cents each, plus shipping, from the Museum Store at the Art Institute of Chicago, (312) 443-3535.

Seurat images may also be available on the Internet. Try conducting a search for some possible URLs using "Seurat" or "pointillism" as keywords.

Special Section on Optics

by **Michael Isaacson, N. S. Kapany Professor of Electrical Engineering and Associate Dean for Research, Baskin School of Engineering, University of California at Santa Cruz.**

One of the most revealing ways to understand how a microscope works is to go through the activities below on your own. The exercise should help you understand how simple lenses work and introduce basic features of microscope design. The GEMS guide *More Than Magnifiers* may be helpful in providing more classroom activities concerning lenses.

For the activities described below, you need the following items:

   ___ *(optional)* a water-filled fish tank, under 10 gallons
   ___ a flashlight (preferably one with a bright, focusable beam)
   ___ a clear, smooth-sided, round glass filled with water
   ___ several large blank white cards
   ___ a comb
   ___ four plastic hand magnifiers (two each of two different focal lengths, about 2" and about 3")
   ___ an index card with typed text or picture
   ___ heavy duty spring clothespins (at least 4)
   ___ a meter or yardstick

Note: Scientific supply companies carry optical bench sets, lenses, lens holders, and related equipment that could be used in creating and experimenting with the microscope described in the last part of this series of activities. For example, Nasco (see the "Sources for Materials" section) has an inexpensive basic optical bench set, along with lenses and other related accessories. Many high school physics labs are likely to have similar equipment.

In order to understand how a microscope works, we must first understand what lenses do to light.

What do lenses do to light?

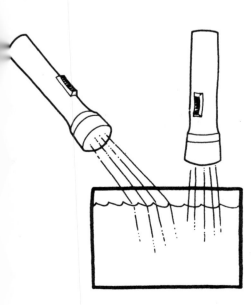

Light generally travels in straight lines. But, if you put any clear material that is more dense than air (water, glass, plastic) at some angle to a light beam, the light will bend upon going into that material from the air. You can observe this with a fish tank full of water by shining a flashlight at an angle to the water surface and looking at the light path from the side.

Notice that if the light beam goes straight into the water (perpendicular to the surface), the light also goes straight into the water. But as the light beam makes a larger and larger angle to the perpendicular, it is bent more and more, at the point where it enters the water.

A lens is a curved piece of clear material that bends light like the water in the fish tank. Lenses are usually made of glass or plastic. The light bends more or less depending upon the curvature (or surface angle) of the part of the lens it strikes. Light entering the center of a lens goes straight through, since it enters perpendicular to the lens surface, but light entering near the edge gets bent inwards towards the center.

If the curve is right, there is a point where all light coming from very far away gets bent or "focused" to a point. This point of focus is called the "focal point" of the lens and the distance of this point from the lens is called the "focal length." Lenses with greater curvature bend the light more and have shorter focal lengths—we say they are more powerful. (If you've gathered the materials to work through this section, you have lenses of two different focal lengths.)

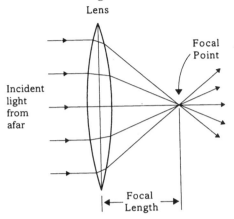

You can make a simple lens with a clear glass of water and can observe the "bending power" of this lens with a comb, in a darkened room. Put the flashlight flat on the table so you can see a beam of light on the table surface. Now put a comb on the front of the flashlight (less than one inch from it) so that the light shines through the teeth of the comb. You now have many light beams. Put the glass of water in front of these beams (close to the comb). You will see light on the opposite side of the glass bend inward towards the center almost forming a point. It's not a "good" point because the water glass is a poor lens. It's a poor lens because it is only curved in one direction, so it doesn't focus or magnify in the perpendicular direction (along the axis of the cylinder). Therefore the image is distorted. A glass ball, such as a clear marble, solves this cylinder problem, and will also work as a lens (as your students discover in the Up Close station), but it has the same optical problem as the glass of water. Rays passing through the periphery are bent so much that they aren't focused at the "focal point." Light that strikes the sphere far from its axis is bent so much that it is focused to a different point than light that strikes the sphere near its axis. The farther away from the axis the worse it gets. In optical terms, this phenomenon is called "spherical error" or "spherical aberration" and it causes blurring.

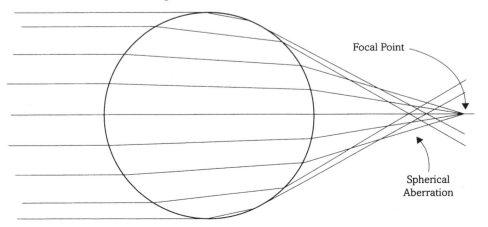

How do we cure this problem and produce a good lens?

Cut out a paper circle to represent a spherical lens. Fold it in half. Draw a line parallel to the fold, about 2/3 of the distance from the fold to the edge of the half circle. Cut on that line and tape the two crescents together along the straight edge. You now have the shape of a good lens. Flatten the large folded piece and look at it. What remains? [The extremely curved edges of the sphere far from its diameter and the area of the sphere in the center.] The area in the center does no useful "optical work" since it functions just like a piece of flat glass. The area at the edges distorts the light that passes through because the sphere is not a perfect focusing curve. A sphere makes a good lens only if we don't use these extreme edges, thus avoiding blurring or "spherical aberration."

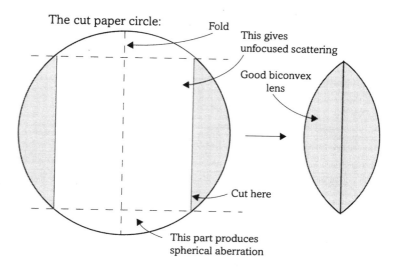

The cut paper circle: Fold

This gives unfocused scattering

Good biconvex lens

Cut here

This part produces spherical aberration

Note: You can actually make lenses with more complicated curvatures than a sphere and get perfect focusing, but they are difficult to design and expensive to make. Such lenses are said to be "corrected" for spherical aberration. In fact, prehistoric creatures called trilobites had eyes with such "perfect" lenses!

Can you think of a way to determine the focal length of a lens?

One good way is to hold your lens in front of a white card and let light from a distant light bulb (such as one on the ceiling) strike the lens. What do you see on the card as you move the lens further away? The point where you see the light bulb "in focus" is the "focal length" of the lens—if the bulb is far enough away. You could also hold the lens far away from a bright window. On the opposite side of the lens from the window, slowly move a white card away from the lens. The distance at which the image of the window scene is "in focus" on the card is the focal length of the lens. Does the focal length change if you turn the lens around?

Now, looking very carefully at the image, what does it look like?

You will note that it is an inverted image (upside down). If an object is at a distance from a lens greater than its focal length, the image produced on the opposite side of the lens is "real" and inverted. A "real" image means that it can be projected onto a card placed at a point along the line of lenses (the "optical axis").

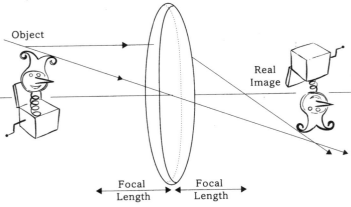

If you try to put the object closer to the lens than its focal length you will get an upright "virtual" image. "Virtual" means that you are not able to project a focused image on a card placed where the image is formed.

To use your lens as a "magnifier," hold it near this page. Determine the distance between the lens and the page for a clear magnified image. Is the image inverted or right side up? How does this distance compare with the focal length of the lens?

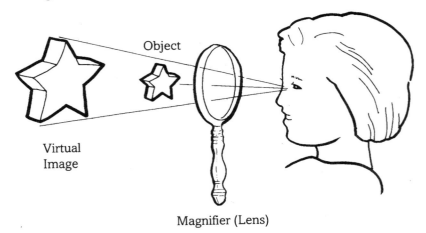

Magnifier (Lens)

The distance away from the page for a clear image is less than the focal length and the image is not "real"— it is "virtual." The light appears to come from the image and it is "magnified." A simple "magnifier" (one lens) produces a virtual, upright image and "fools" you into thinking the light came from a larger object. (See "Behind the Scenes" for a brief explanation of why we perceive the object as magnified.)

The lenses called for in the activities below are of two different focal lengths. Determine the focal length of each. (A word of caution—the focal length is related to the **curvature of the lens**, *not* the diameter, even though for these particular lenses the one with the larger focal length also happens to be larger in diameter.)

A compound microscope uses at least two lenses. Why do you think this is desirable? This next activity will allow you to construct such a microscope.

Construction of a Compound Microscope

First, mount a card with text, or an image, using a clothespin as shown below. The clothespin should lay flat on the table with the card sticking up. You need to have good illumination of the card (your "object") so that there is enough light reflecting off it into a lens to form an image that you can easily see (a good flashlight will do). It is best not to use reflecting or shiny objects since this makes it more difficult to get the illumination even and it becomes harder to set up the microscope.

Take the lens with the longer focal length and mount it on another clothespin. Start out with the lens about 4" from the object card. Use a blank card mounted on a third clothespin to locate the image. Is it further from the lens than the object was? Is it a magnified image? Is it inverted or right side up?

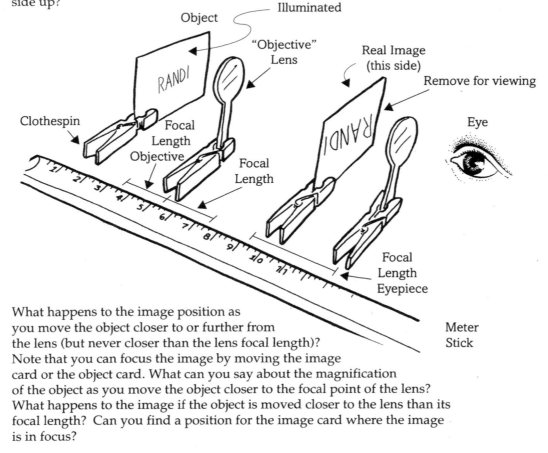

What happens to the image position as you move the object closer to or further from the lens (but never closer than the lens focal length)? Note that you can focus the image by moving the image card or the object card. What can you say about the magnification of the object as you move the object closer to the focal point of the lens? What happens to the image if the object is moved closer to the lens than its focal length? Can you find a position for the image card where the image is in focus?

Now go back to having the object card about 4" from the lens. This first lens is called the "objective" lens because it is close to the object. The image produced is called the intermediate image of the microscope you are building—it is magnified, real, and inverted. Take a second lens and mount it as before on a clothespin. For starters, choose the same strength lens as the objective. Position it so that the intermediate image is closer to this second lens than its focal length. This second lens is called the "eyepiece" since we will be looking through it to see our final image. Now remove the card at the position of the intermediate image and look through the eyepiece. This intermediate image is now being magnified by the eyepiece in the same way as a simple magnifier. Be sure the two lenses are aligned with one another so their centers are on the same line (the optic axis) or you may get severe distortion and won't be able to easily see a magnified image of the object. Look through the eyepiece and slightly adjust its position and the objective lens position to see a magnified inverted image of the object card. Is this a "real" or "virtual" image?

You have now built a compound microscope. It is instructive to see which of the two lenses that you are using (objective or eyepiece) is best used for focusing. Can you focus by only moving the object?

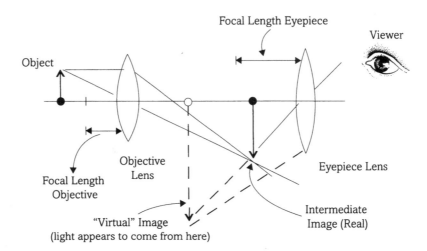

Generally, in a microscope, the objective lens has a **shorter** focal length than the eyepiece. The shorter the objective focal length, the larger the magnification. A short focal length objective is a little trickier to focus. Now that you have developed your microscope skills, you may want to try using the shorter focal length lens as the objective (the one with a focal length around 2"). Now go through the entire exercise again. (You'll need to find the new intermediate image again, and proceed as above.) In a real microscope, the intermediate image is formed in roughly the same plane so that the microscope length doesn't change as we change objective lenses. But in this microscope exercise, that is not the case.

The magnifying power of the microscope depends upon the product of the power of the objective lens (which gives you the intermediate image magnification) times the eyepiece (which acts as a simple magnifier). This is why microscopes have two basic lens systems; objective and eyepiece. It isn't possible to achieve high magnification with just one.

Can you think of a way you could make a microscope that would give an "upright, non-inverted image?" That is precisely what the optical system used in an inspection/dissection microscope does. See the "Behind the Scenes" and "Resources" sections for more information and for resource books on home-made and/or inexpensive microscopes and lots more about how microscopes work!

Resources

The listing below is divided into five main sections: Curriculum and Activity Guides; Resource Books; Videotapes; CD-ROMs; Internet Connections; and a Science Center Exhibition. The listing is condensed, modified, and adapted from a more complete bibliographic listing, with annotations, by Caroline Schooley. That complete list is available on the Internet at: **http://www.msa.microscopy.com/ProjectMicro/**

The materials in this list cover a wide range of prices. The full list on the Internet includes price information.

Curriculum and Activity Guides

Anderson, R. and Druger, M., eds., 1997, **Explore the World Using Protozoa**, order from the National Science Teachers Association, 1840 Wilson Blvd., Arlington, VA 22201-3000; 800-722-NSTA. NSTA collaborated with the Society of Protozoologists to produce this reviewed collection of 28 investigations; microscopes are the only "specialty equipment" required. Shows how to present an entire "live" biology class (morphology, physiology, ecology, ethology, taxonomy—everything!) with protozoa. It's intended for grades 9 and up, but parts can be adapted for middle school. More information available at: http://www.nsta.org.pubs.

British Nuclear Fuels plc, 1995, **The Young Detectives,** BNFL Education Unit, P.O. Box 10, Wetherby, West Yorkshire LS23 7EL, England. The Royal Microscopical Society sponsored the development of this teaching pack, which comes with a teachers' manual, photocards, wall posters, a 20-minute video, and a computer disk. The mystery is a case of apparent vandalism at Sherlock School. Grades K-6. **(Out of Print)**

Delta Education, 1987, **Small Things: An Introduction to the Microscopic World**, Delta Education, Inc., P.O. Box 950, Hudson, NH 03061-6012; 800-442-5444. This teacher's guide, a series of 11 lessons, is part of the Elementary Science Study series. The first unit introduces lenses and microscopes; the others feature crystals and biology. Grades 4–6.

Discovery Scope, Inc., 1992, **Activity Booklets**, Discovery Scope, Inc., 3202 Echo Mountain Dr., Kingswood, TX 77345; 800-398-5404. Titles available in this series are: **Investigating wetlands, Investigating arthropods, Investigating seashore life, Investigating termites, Investigating protozoans**, and **Macrophotography.** All emphasize the Discovery Scope and its accessories. The photography booklet is particularly useful for the teacher or advanced student who wants to do still or video photos. Middle school–adult.

Hixson, B.K. and Hutson, T.L., 1993, **Microscopes**, order from the Wild Goose Co., Salt Lake City, UT 84115; 801-466-1172 or from Edmund Scientific. A comprehensive three week course that teaches care and use of a microscope, identification of parts, optics, and several slide preparation techniques. Includes tests, puzzles, and humorous illustrations. Grades 4–adult.

Insights Visual Productions, 1991, **Experiments With the Mini-Microscope: Teacher Handbook**, order from Insights Visual Productions, Inc., P.O. Box 230644, Encinitas, CA 92024; 619-942-0528 or 800-942-0528. Seven classroom sessions, with advice on lab preparation and procedure, background information, student worksheets, overhead masters, and quizzes. Topics include use of a 30x handheld microscope, measurement, and investigative observation of a variety of objects. Comes with a 50 minute video (see below). Middle school.

Museum of Science, Boston, 1997, **Magnification and Microscopes**, Science Park, Boston, MA 02114-1099. The manual is included with a rental kit for a class of 30. The kit includes 15 hand lenses, 15 handheld 30x microscopes, one 25x–100x compound microscope, and supplies for five weeks of microscopy. The manual includes sequenced lessons on all the "basic" subjects—such as crystals, fingerprints, pond life, sand, and color printing—as well as assessment suggestions and copyable student sheets. Middle school.

NSRC, 1991, **Microworlds**, National Science Resources Center, Smithsonian Institution/National Academy of Sciences, Arts and Industries Bldg., Room 1201, Washington, DC 20560; or order from Carolina Biological Supply. This unit, from the Science and Technology for Children series, contains 16 lessons on lenses, use of a 30x microscope, field of view, halftone illustration, and observation of a variety of specimens. The **Discovery Deck** is a 1997 addition to Microworlds. It's a boxed set of 30 double-sided cards on microscopes and microorganisms with classroom activities. Grades 4–6.

Ruef, K., 1992, **The Private Eye,** The Private Eye Project, P.O. Box 646, Lyle, WA 98635; 509-365-3007. This manual uses a 5x jeweler's loupe to construct an integrated curriculum, over the full K-12 grade range. The central role of imagery in science is used to link science to visual and verbal arts, and magnification is used as the key to observation. Study of simple "found" objects is followed by descriptive analogy and development of theory. Introduces creative use of inquiry for teachers and students and can help science professionals learn how to present their specialty in an exciting way that encourages further inquiry. More information available at http://www.the-private-eye.com/ruef/ Grades K-12 and adult.

Sherer, M. and Isenberg, S., 1991, **Microscope Study**, order from Putnam/North Westchester BOCES, Yorktown Heights, NY 10598; 914-245-2700, x349. This New York State Elementary Science Program manual contains 20 class exercises—a semester's worth of activities. Inquiry science methods introduce hand lens and microscope use, slide preparation, and a wide variety of "standard" biological and physical science subjects (onion cells, yeast, pond water, crystals, fingerprints, etc.). Grades 5–6. (out of print)

Zook, D. and "the Microcosmos Team," 1992, **The Microcosmos Guide to Exploring Microbial Space**, Kendall Hunt Publ. Co., 2460 Kerper Blvd., Dubuque, IA 52001; 800-338-5578. Focused on microbiology, this can be used as a complete curriculum or it can be broken into stand-alone units. Two units—*Magnifying without money: easy classroom activities with inexpensive microscopes* and *The microdiscovery board: approaches using the 30x microviewer*—emphasize microscopical method. Grades 7–9. (out of print)

Resource Books

Burgess, J., Marten, M., and Taylor, R., 1987, **Microcosmos**, Cambridge University Press, New York or order from Carolina Biological Supply. In this beautiful volume, light and electron micrographs are presented in color and black and white; non-biological subjects occupy a third of the book. The 15-page "technical appendix" is an excellent concise description of the varieties of modern microscopy. Middle school–adult; the images are useful for all ages.

Canault, N., 1993, **Incredibly Small**, New Discovery Books, Macmillan, New York. An excellent SEM picture book, with strong educational content. (SEM refers to scanning electron micrographs—images obtained with a scanning electron microscope.) The pseudocoloring is explained, and "real life" colors are used. Thirty-six everyday objects are presented (salt and sugar, dust, dog hair, polystyrene, pollen) and placed in context with unmagnified photos and text. All ages.

Cobb, V. and J., 1993, **Light Action!**, HarperCollins, New York. Subtitled "amazing experiments with optics," this manual includes easy-to-do experiments using common home supplies. Optical concepts are explained well. Specific information on lenses is in one chapter. Advanced topics such as polarization and diffraction are included, which will help young crystal-growing enthusiasts. Grades 1–high school.

DaMert Co., 1996, **Discover Magnification**, contact the manufacturer, DaMert, 1609 Fourth St., Berkeley, CA 94710; 510-524-7400, for bulk orders and retail sources. DaMert has packaged a magnifying "bug box," a plastic folding triplet hand lens, and a cased Fresnel lens (with inch and centimeter scales) with a concise description of refraction and lenses. Although it isn't a "book," it is carried by some bookstores. Grades 3–high school.

Edwards, F.B. and Aziz, L., 1992, **Closeup: Microscopic Photographs of Everyday Stuff**, Firefly Books, P.O. Box 1338, Elliott Station, Buffalo, NY 14205. Twenty-one common objects are presented as "guess what this is?" scanning electron micrographs. It's an interesting assortment, including popcorn, candy, Velcro, and polystyrene foam, as well as insects and flowers. All ages.

Headstrom, R., 1941, 1977, **Adventures with a Microscope**, Dover Publications, Inc., New York or from Delta or Carolina Biological. An excellent observational exercise collection; the wide variety of its 59 "adventures" allows a teacher to use available materials and advanced students can use independently. High school–adult. There are two other books for adult learners by the same author, **Nature Discoveries with a Hand Lens** and **Adventures with a Hand Lens**.

Levine, S. and Johnstone, L., 1996, **The Microscope Book**, Sterling Publishing Co., New York. Excellent general introduction to microscopes and microscopy, with more than 36 exercises. A third of the book is devoted to optical principles, including field of view, measurement, and depth of focus. Another third is biology (including making a hand microtome and a well slide), and eight exercises present geology, crystal growing, forensics, and air pollutants. Many line drawings and color photomicrographs. Middle school.

Loewer, P., 1996, **Pond Water Zoo**, Athenium Books for Young Readers, Simon & Schuster, New York. This is a natural history of commonly encountered pond life, from monera to micro-arthropods. The illustrations are excellent black and white drawings, and it's written for young readers without being simplistic. Middle school.

Lovett, S., 1993, **Extremely Weird Micro Monsters**, John Muir Publications, P.O. Box 613, Santa Fe, NM 87504. A variety of microorganisms, from viruses to insects, are covered; the images are pseudocolored. They've been selected for visual impact; most (e.g., tapeworm, tardigrade) won't be seen by the average student, but they're delightfully ugly. Middle school–adult.

Molitor, D.L., 1983, reprinted 1991, **Fun at Forty Power; Microscope Projects for Beginners,** The Lab Bench, 21 Morningside Dr., St. Paul, MN 55119-5006; 612-730-7184. An inexpensive, useful pamphlet which includes a brief description of preparation methods for a variety of specimens including unique suggestions for preparation of surface replicas using drugstore supplies and section stain from a tropical fish store. Grades 6–adult.

Morrison, P. and Morrison, P., 1982, **Powers of Ten**, A Scientific American Library book form W. H. Freeman & Co., New York. Forty two different views of the same picnic are shown—each one a power of ten different than the view it borders. The views begin from about 1 billion light years away from earth, travel through our familiar daily scale, then go on to the microscopic. The book was created after the success of a 16mm movie (now a video; see below). Written for adults—a useful teacher reference for use with the video. Now also available as a "flipbook" of the images, from the galaxy to a proton. Middle school-adult.

Murphy, P. and the Exploratorium, 1993, **Bending Light**, Little, Brown & Co., Boston. Only two pages focus on microscopes, but this is an excellent introduction to properties of lenses. Convex and concave lenses, focal length, and refraction are presented clearly. Includes "dozens of activities for hands-on learning." Comes with an inexpensive plastic lens. Middle school.

Nachtigall, W., 1995, **Exploring With the Microscope**, Sterling Publishing Co., New York. Intended for adult amateur microscopists, this book could provide teachers and classroom volunteers with useful information on "serious" light microscopy. Almost half of the book is devoted to simple preparation methods for biological specimens and descriptions (with good illustrations) of commonly encountered organisms. Adult.

Norden, B. and Ruschak, L., 1993, **Magnification**, Lodestar Books, 575 Hudson St., New York, NY 10014. This eye-catching pop-up book presents artificially colored scanning electron micrographs of common objects, at various magnifications. All ages visually, but the brief text is adult.

Ontario Science Center, 1987, **Have Fun With Magnifying**, Kids Can Press Ltd., 5855 Bloor St. West, Toronto, Ontario, Canada M6G 1K5 or order from Delta Education. This picture book presents water-filled vials as cylindrical lenses. Comes with 30 clear plastic vials. Grades 1–3.

Oxlade, C. and Stockley, C., 1989, **The World of the Microscope**, Usborne Publishing Ltd.; in the U.S., from EDC Publishing, 10302 E. 55th Pl., Tulsa, OK 74146 or from Cuisenaire. An illustrated do-it-yourself student manual. The emphasis is on observation, with both biological and physical science projects. Use of a student compound microscope and simple specimen preparation methods are described. Optics discussed briefly. Grades 6–9.

Rainis, K.G., 1991, **Exploring With a Magnifying Glass**, Franklin Watts, Danbury, CT. This book begins with a good optics chapter with instructions for two home-built simple microscopes. Several "nature study" chapters follow, with questions that could become science projects. Material on photography, metals, paper, and fabric included. Middle and high school.

Rainis, K.G. and Russell, B.J., 1996, **Guide to Microlife**, Franklin Watts, Danbury, CT. This is a comprehensive microworld field guide with many color light micrographs. The authors state the 115 microorganisms described comprise 75–90% of those that may be encountered in the "wild." Habitats described are diverse, with advice on collecting methods. Organisms are also diverse, from monerans to millimeter-sized arthropods. Species descriptions include ecological information, collection, culture, and ideas for further investigation. Middle school–adult.

Rogers, K., 1999, **The Usborne Complete Book of the Microscope**, Usborne Publishing Co. Ltd./EDC Publishing, Tulsa, OK. Usborne books have an earned reputation for good design and spectacular illustrations; this one is no exception. It's full of accurate, well-written information on microscopes, simple preparation methods, and the microworld. Subjects such as common objects, crystals, the human body, bugs, rocks, and simple nanotechnology will interest the young beginner. It concludes with a bit of history, advice on buying a first microscope, a glossary, and a good index. Middle and high school.

Ross, M.E., 1993, **The World of Small: Nature Explorations With a Hand Lens**, from the Yosemite Association, Yosemite National Park, CA; 209-379-2648. A well-written, nicely illustrated, rugged (it's printed on card stock) introduction to "nature explorations with a hand lens." Included with the book is a quality 5x lens. Middle school.

Shih, G. and Kessel, R., 1982, **Living Images**, Science Books International, 20 Providence St., Boston, MA 02116, or order from Wards. This excellent collection includes over 350 micrographs of a wide range of living things. The brief text doesn't assume advanced knowledge. An excellent classroom resource. Grades 6–adult.

Stidworthy, J., 1993, **Insects Through the Microscope**, Shooting Star Press, 230 Fifth Ave., New York, NY 10001. Text and microscopic photographs introduce the physiology, reproduction, growth, and metamorphosis of insects. Several pseudocolored light or electron micrographs per page make this an eye-catching book. Part of the Through the Microscope series. Grades 8–adult.

Tomb, H. and Kunkel, D., 1993, **MicroAliens: Dazzling Journeys with an Electron Microscope**, Books for Young Readers, Farrar, Straus & Giroux, New York. A good presentation of the SEM microworld for young people. It begins with a two-page non-technical description of electron microscopes, then explores the everyday worlds of air, water, yard, home, "you," and "inside you." Includes a comparison of hooks on a seed with Velcro and a sidebar on Velcro's invention. Middle school–adult.

Tomecek, S., 1995, **Bouncing and Bending Light**, Scientific American Books for Young Readers, W.H. Freeman & Co., New York. The *Phantastic Physical Phenomena* book series parallels the Dr. Dad's PH3 video series (see below). Presents reflection, refraction, and basic lens theory with experiments, analogies, and stories of great discoveries. Inquiry is stressed, and answers to text questions are given in mirror writing, which is likely to appeal to students. Useful with or without the video. Grades 4–8.

VanCleve, J., 1993, **Microscopes and Magnifying Lenses**, John Wiley, New York or order from Delta. A collection of 20 interesting projects. The first seven exercises introduce simple and compound lenses. Most of the rest use a simple magnifier with less emphasis on a compound microscope. Grades 6 and up.

Wood, R., 1990, **Physics for Kids: 49 Easy Experiments with Optics**, McGraw-Hill/Tab Books, Blue Ridge Summit, PA or order from Edmund Scientific. A collection of 30-minute projects, requiring only common household items. Grades 3–7.

Zubrowski, B., 1992, **Mirrors: Finding Out About the Properties of Light**, Beech Books, Morrow & Co., 1350 Avenue of the Americas, New York, NY 10019; 800-282-8257 or order from Cuisinaire. This Boston Children's Museum Activity Book is filled with activities that contribute to understanding the properties of light. Suitable for class or independent use. Grades K–6.

Videotapes

All are VHS format. Many videos are relatively expensive; if you're hesitant about a purchase, ask the supplier if a preview copy is available.

Acorn, J., 1994, **Under the Microscope**, 24 minutes. Produced by CFRN Television and Great North Productions, Edmonton, Alberta, Canada. Order from Filmwest Associates, 300 W. 2nd St., Carson City, NV 89703; 775-883-8090; e-mail: info@filmwest.com. This is tape #6 in the Acorn, the Nature Nut series. Using humor, role-playing (van Leeuwenhoek and a science nerd), and song, this tape provides advice on how to select an inexpensive compound microscope and study pond life. For middle school, plus the rest of the family.

BioMedia Associates, 1992, **Exploring with a Hand Held Microscope**, 18 minutes. Order from BioMedia Associates, P.O. Box 457, Loomis, CA 95650; 916-663-3304. This is an introduction to the collection and study of pond organisms, intended for Discovery Scope users (see Curriculum and Activity Guides above).

BioMedia Associates, 1992, **Imaging a Hidden World**, 15 minutes. Order from BioMedia Associates, P.O. Box 457, Loomis, CA 95650; 916-663-3304 or Discovery Scope, Inc., 3202 Echo Mountain Dr., Kingswood, TX 77345; 800-398-5404. Videomicroscopy with a home camcorder is demonstrated. Accessories from Wards and Discovery Scope are emphasized, but the information could be adapted to other equipment. Includes an introduction to the optics and operation of a compound microscope. Middle school–adult.

BioMedia Associates, 1999. **Eye of the Cyclops** Eight 15-minute tapes, also available separately from BioMedia, 888-248-6665, www.biovideo.com, Connecticut Valley Biological Supply, 800-628-7748, or Carolina Biological, 800-334-5551. Available in both VHS and PAL. The best available video introduction to microlife for middle school. Clever "aquanaut" adventures stress observation, inquiry, and experimentation and explore many topics. The point is clearly made that there's still much to learn about microlife. The photography is superb. The segments are short enough to use in a class hour that includes microscope observation and discussion. Study guides with student worksheets are available for each segment. The tapes are: Plankton Play, Decomposers Everywhere!, Food Chains Begin with Photosynthesis, Protozoans and Algae, Predators of the Shallows, White Water Adventure, Discovering a Forest Microcosm, and Backyard Biodiversity. Middle school.

Childrens Television Workshop, 1991, **Crystals: They're Habit Forming**, 15 minutes. Order #2 in the 3-2-1 Classroom Contact series from GPN, P.O. Box 80669, Lincoln, NE 68501-0669; 800-228-4630. The formation of a crystal lattice is illustrated with an animation sequence and molecular models. Snow crystal growth is discussed and demonstrated. Grades 4–6.

Childrens Television Workshop, 1991, **Refraction: Facts of Light**, 15 minutes. Order #25 in the 3-2-1 Classroom Contact series from GPN, P.O. Box 80669, Lincoln, NE 68501-0669; 800-228-4630. An engaging middle-school-student narrator, plus use of both animation and a laser beam make this an excellent introduction to refraction and convex lenses. Grades 4–6.

Colgren, J., 1989, **How to Use a Microscope**, 20 minutes. Order from United Learning, Inc., 6633 W. Howard St., Niles, IL 60714-3389; 800-424-0362 or from Nasco. A monocular compound microscope with mirror and no condenser is shown, which may help the teacher who has that equipment. Instructions are clear and brief. About half the tape includes "other microscopies," mainly SEM images of common objects. Includes three-page teacher's guide and student sheets.

Delta Education, 1994, **Lenses and Mirrors**, 18 minutes. Order from Delta Education, Inc., P.O. Box 3000, Nashua, NH 03061-9913; 800-282-9560. This is divided into three parts: mirrors and reflection, lenses and refraction, and lenses and mirrors in action. Grades 5–6.

Eames, C. and Eames, R., 1978, **Powers of Ten**, 21 minutes. Pyramid Film & Video, 2801 Colorado Ave., Santa Monica, CA 90404; 800-421-2304. Also available as #A1546 from PBS Home Video, 800-645-4PBS or from Arbor Scientific, P.O. Box 2750, Ann Arbor, MI 48106-2750; 800-367-6695. This classic places the microworld in the size continuum; it goes from galaxies to atomic nuclei by 10x steps. Middle school–adult.

Great Plains National, 1988, **Germs Make Me Sick!**, 30 minutes. Order as #E126-34 in the Reading Rainbow series from GPN, P.O. Box 80669, Lincoln, NE 68501-0669; 800-228-4630. Microscopy is emphasized, but it's primarily an introduction to microbiology. A brief sequence shows pond water collection and observation, followed by microscopy of common bacteria. Much of the tape is a reading of the children's book that shares the title. There is a two-page teacher's guide. Grades 3–6.

Hatch, Warren A. Video Productions. Warren Hatch is a schoolteacher who makes videotapes related to microscopy. Common objects are shown at from 10x–100x, and sequences are time-listed for easy reference. He has produced a video to accompany *Microscopic Explorations*. It has video footage related to each of the learning station activities, and is entitled

Video for Microscopic Explorations, 1998, 30 minutes. This tape and the others listed below can be ordered from Warren Hatch, P.O. Box 9224, Portland, OR 97207; e-mail: whatch@hevanet.com. Hatch's other videos include: **Video Microscope Short Highlights,** 1996, 32 minutes, a "greatest hits" collection—good introduction to classroom microscopy; **Video Microscope I,** revised, 1993, 120 minutes, sand grains from around the world, a burger and fries; **Video Microscope II,** 1992, 120 minutes, spiders and insects, fabrics, zippers, Velcro, 38 minutes of pond microorganisms, foods (including pizza and potato chips), microcrystals, color printing, and seeds; **Video Microscope III,** 1993, 121 minutes; **Video Microscope IV,** 1996, 41 minutes; **Pond and Puddle Life through a Microscope,** 1993, 121 minutes; **Protozoa through a Microscope,** 1996, 90 minutes; **Los Angeles River through a Microscope,** 1994, 84 minutes, emphasis on insects; **Spiders and Mites through a Microscope,** 1993, 89 minutes; **Insects and Spiders Up Close,** 1996, 37 minutes, 16 common arthopods—for kindergarten and first grade; **Sand through a Microscope,** 2nd edition, 1996, 52 minutes; **Seeds through a Microscope,** 2nd edition, 1996, 54 minutes; **Crystals through a Microscope,** 1994, 81 minutes, 38 different chemical crystals shown growing in real time; **What's That through the Microscope?**, 1994, **#1,** 50 minutes; **#2,** 60 minutes; **#3,** 53 minutes, built around a question format; **Spiders and Mites Alive through the Microscope,** 1997, 30 minutes; **Worm Bin Creatures Alive through a Microscope,** 1998, 32 minutes; **Crystal Growing: Views through a Microscope,** 1997, 30 minutes; and **Protozoa in Action,** 1997, 30 minutes.

Insights Visual Productions, 1991, **Experiments With the Mini-Microscope**, 50 minutes. Insights Visual Productions, Inc., P.O. Box 230644, Encinitas, CA 92024; 619-942-0528 or 800-942-0528. A wide variety of common objects (money, fingerprints, fabric, etc.) observed with a 30x handheld scope. In-depth treatment designed to provoke further inquiry. Calibration and measurement are introduced well and used throughout. The teacher's manual is described in the Curriculum section above. Middle school. **Experiments With the Microscope,** 1992, 66 minutes. Narrator/producer/ex-teacher Jack Ross extends the inquiry begun at 30x in the previous tape to compound microscope magnifications. Includes an introduction to the care and use of the microscope, how to make various types of slides, and how to measure specimens (an important and often ignored topic). Middle and high school **How To Use the Microscope,** 1994, 12 minutes. This tape is identical to part one of Experiments With the Microscope, listed above. An introduction to the compound microscope and slide making. Middle and high school. **Smaller than the Eye**

Can See. 1989, 14 minutes. Free-living protists and microscopic animals are introduced with good photography; the narration and teacher's manual assume no previous knowledge. All ages.

Louisiana Public Broadcasting, 1994, **Optics: Beginning to See the Light,** 15 minutes. Order #602-8 in the Dr. Dad's PH3 series from GPN, P.O. Box 80669, Lincoln, NE 68501-0669; 800-228-4630. Dr. Dad uses simple experiments with a flashlight and glasses of water to demonstrate refraction and the focal point of lens to a group of students. One of the students then goes for an eye exam which nicely relates lenses to vision. Prisms and a Fresnel lens are introduced. The four page teacher's guide has both experiments and background information. Grades 3-6.

Matulavich, P., 1989, **Lenses and Magnification,** 20 minutes. Produced by Barr Films and distributed as #A556 by Clearvue/eav, 6465 North Avondale, Chicago, IL 60631-1909; 312-775-9433 or 800-253-2788. A good introduction to optics, beginning with an explanation of refraction, then magnification, reduction, focal point, and focal length. These concepts are then applied to compound microscopes and telescopes. The short segments on scanning electron microscopy and radio telescopes could confuse the introduction to optics. Middle school–adult.

Matulavich, P., 1987, **The Microscope and Its Incredible World,** 21 minutes. From Barr Films; distributed as #A432 from Clearvue/eav, 6465 North Avondale, Chicago, IL 60631-1909; 312-775-9433 or 800-253-2788. Explains the contribution of objective and eyepiece to microscope magnification and provides good advice on using a student microscope. The visual effect of changing focus is demonstrated well, and brightfield and darkfield are mentioned but not explained. A variety of nicely photographed specimens follow. Middle school–adult.

Sitko, J., 1991, **Introduction to the Microscope.** This two-tape series is #CL317-CV from Clearvue/eav. It includes #CL318-CV, **How to Use the Compound Microscope,** 20 minutes, and #CL319-CV, **The Hidden World of Microscopy,** 19 minutes. The first tape begins with history, then discusses the parts, care, and use of a student microscope. The second tape features photomicrography of living microorganisms. Gives advice on preparing hay infusions and wet mounts. Middle school-adult.

WGBH, Boston (NOVA), 1996, **Odyssey of Life,** 3 parts, 60 minutes each. WGBH Boston Video, P.O. Box 2284, South Burlington, VT 05407; 800-255-9424. A three part series which features the microphotographic work of Swedish photographer Lennart Nilsson, ranging from human reproduction to dust mites to microscopic viruses. There's also information on Nilsson and how he accomplishes his amazing photography. All ages.

WTTW, Chicago and Kurtis Productions, 1992, **Mystery Through the Lens,** 30 minutes. Order from The New Explorers, Public Media Education, 5547 N. Ravenswood Ave., Chicago, IL 60640-1199; 800-343-4312, x359. The series focuses on the research of outstanding scientists and this episode features light microscopist Walter McCrone and his detective work, including his analysis of the Shroud of Turin. Good support to a forensics unit. Grades 7–12.

Please see the more complete and updated listing of CD-ROMS on the Project Micro website under Resources for many more CD-ROMs.
(www.msa.microscopy.org/ProjectMicro)

CD-ROMs

Many suppliers will allow schools to preview CD-ROMs before purchase.

Baggott, L., 1999?, **Interactive Microscope Laboratory,** published by Addison Wesley Longman (Interactive Learning). Over 200 samples are included on this CD-ROM. Focus and magnification are button-controlled, and there are quantitation tools. School-tested in the United Kingdom. See the web site http://www.ex.ac.uk/telematics/IML. Grades 9–11.

Corel Corporation, **Corel Professional Photos**, contact Corel, 1600 Carling Ave., Ottawa, Ontario K1Z 8R7, Canada or call 613-728-8200 for other dealers. For both Mac and Windows. Copyable professional stock photos are an excellent source of images for classroom use. Corel has hundreds of titles; two relevant ones (each with 100 images) are #645000, **Microscopic Images** (colorized SEMs of biological subjects) and #610000, **Sand Grains of the World** (stunning images of shell, gemstone, and other mineral sands). Both computer formats provide thumbnail views, but the CD itself provides a "slide show" only for Windows; Mac users must use an image program to display full screen photos. Adult.

Corel CD Home, 1997, **Beyond the Naked Eye**, from the Edutainment Catalog as item #353049; call 800-338-3844 for other retail sources. For DOS, Windows 3.1, and Mac. This three-part Life Science Mystery series begins with the microscopic world of bacteria and viruses. Two situations are presented to the student-scientist: diagnosis of a bacterial disease and creation of an AIDS-awareness program. There's a lab in the research center, with an incubator (for the bacteria), light and scanning electron microscopes, and a harassed technician. The microscopes are used for these microbiological purposes and scientific method is presented well. The second CD is **The Green Files** and the third is **Crisis at the Anamalia Research Center**. **Beyond the Naked Eye** is similar to **Science Sleuths** (see below), but for somewhat older students; grades 8–12.

Michigan State University, Communication Technology Lab and Center for Microbial Ecology, 1996–97, **Microbe Zoo**, issued by Peregrine Publishers, available from The PC Zone, 800-258-2088, or The Mac Zone, 800-248-0800. Requires recent Mac or Windows platform; check before you order. This CD presents 140 organisms in full-screen micrographs (LM, TEM, and SEM; all monochrome). Students explore the microbe world (bacteria, some fungi, protists, multicellular organisms) by selecting "zoo" habitats (dirt, water, the insides of humans/animals). The "Snack Shop," for example, features microbes that live in snack foods. Within each habitat, students zoom down to see the microbe. The "Size Machine" presents measurement and relative size well. There are four lesson plans, searchable database, and report-writing capability. Preview available at http://commtechlab.msu.edu/ctlprojects/dlc-me/ Grades 5–7; usable in high school.

Microscopy-UK, 1997, **Microscopy-UK Release 1**, available from Molcol Software, c/o Micscape Magazine, 69 Commonside West, Mitcham, Surrey CR4 4HB, UK. For PC only. A "virtual microscope" is presented as a control panel that permits adjustment of many parameters. There are many slide sets (all but one are biological) supplemented with a library of images, QuickTime movies, and articles selected from the popular English amateur microscopy web site. The CD-ROM has won awards in England, and the "virtual microscope" can be downloaded free at http://www.microscopy-uk.org.uk/prodir/software/softmol.html High school–adult.

Neuronware, Scopemaster, 15 Madison Ave, Toronto, Ontario M5R 2F2, Canada. For Mac or Windows. Cannot be ordered direct from Canada. U.S. sources: Clearvue, 800-253-2788, Flinn Scientific, 800-452-1261, and Sargent Welch 800-727-4368. An interactive microscope teaches use of the controls of a compound microscope—the user can select three objectives, adjust the substage diaphragm, use coarse and fine focus. Advice appears if a mistake is made; if advice is ignored, the high power objective even "breaks" with a resounding "crack" if the slide hits it! Ten sets of nine slides each (mostly biological) included, each with reference book. Images are color light micrographs that can be viewed full-screen or reviewed quickly in "teacher" mode. Includes a self and class test. More information at www.neuronware.com Middle and high school.

Schoolhouse Interactive, 1995, **Get Up Close**, published by Schoolhouse Interactive, Inc., Lake Road, Charlotte, VT 05445. Also available from Williamson Publishing Co., 800-234-8791. For either Mac (IIci or above) or Windows. Microscopes and telescopes share this CD-ROM. It's animated, interactive, and accurate. Games and quizzes are used to reinforce, and many light and electron micrographs (including stereo!) can be selected. Quizzes and experiments are middle school level; cartoon antics may appeal to younger students.

Sullivan, J., 1999, **Cells Alive!,** produced by Quill Graphics, 568 Taylor's Gap Rd., Charlottesville, VA 22903, 804-296-8994; order from Yahoo at http://st13.yahoo.net/cellsalive/ or www.cellsalive.com/ For Windows and Mac; requires an Internet browser (Netscape or equivalent). The images from the outstanding *Cells Alive!* web site (www.cellsalive.com/) have been made available in both VHS video and CD-ROM formats, which will greatly increase their classroom usefulness. It's a true copy of the web site, and acts like one—complete with "download" delays. Nevertheless the micrographs, drawings, text and excellent film clips of a variety of living, moving cells and microorganisms will make it a valuable supplement to textbook content. Middle school–adult.

Vazzana, J., 1995, **Microscope—Nature Explorer**, Orange Cherry/New Media Schoolhouse, P.O. Box 390, Pound Ridge, NY 10576; 914-764-4104 or 800-672-6002. The Mac (7.0 or above) version is available now; inquire about Windows. Professor Scope introduces light and electron microscopes in the Nature Center; the menu offers 34 objects to view in the categories air, grass/trees, pond, and rocks. Each object requires mouse "focusing" and is accompanied by written and voice text. Many objects are line drawings, not micrographs, and there is no magnification information. Text is informative, but somewhat oversimplified. Grades 3 and up.

Videodiscovery, Inc., 1995, **Science Sleuths**, Volumes 1 & 2, For Macintosh 7.1 or Windows 3.1. Science Sleuths is an award-winning series of mysteries for middle school students. Designed to build inquiry, critical-thinking, and problem-solving skills, the series challenges students to solve real-world problems by researching clues and conducting science experiments with interactive tools. Each volume has two mysteries, each with 6 solutions of increasing difficulty. Although use of magnification is limited, one mystery in each volume provides a microscope as one of the investigative tools. Includes teacher's manual and assessment software on 3.5" disk. Grades 5–9.

WorldWideWeb

Here's a beginner's list of web sites for K–12 microscopy education.
The first three include hotlinks that will lead you to other excellent pages.
Many are large sites with much more content than the brief category
suggests. This list is kept current in the ProjectMicro bibliography
(http://www.msa.microscopy.com/ProjectMicro),
and is hotlinked there.

K–12 microscopy resources
      http://www.mwrn.com/feature/educatio.htm
"Virtual microscopy library"
      http://www.ou.edu/research/electron/www-vl/
Image hotlinks
      http://www.ou.edu/research/electron/www-vl/image.shtml
Visual and curriculum resources
      http://www.asmusa.org/edusrc/library/
Remote access SEM
      http://www.msa.microscopy.com/MicroScape/MicroScape.html
      http://www.sci.sdsu.edu/emfacility/CUCMEoutreach.html
      http://monarch.mcs.csuhayward.edu/~irsa
Ask a microscopist
      http://www.msa.microscopy.com/Ask-A-Microscopist.html
Image galleries
      http://resolution.umn.edu/MMS/ProjectMicro//gallery.html
      http://www.feic.com/gallery/gallery.htm
      http://microscopy.fsu.edu/galleria/index.html
      http://www.uq.oz.au/nanoworld/images_1.html
SEM image galleries
      http://www.pbrc.hawaii.edu/~kunkel/
      http://www.hitachi.co.jp/Div/keisokuki/english/nano/arte.html
      http://www.pbrc.hawaii.edu/bemf/microangela/
      http://www.mos.org/sln/SEM/gallery.html
      http://www.gatan.com/scharf/scharf.html
      http://www.microscopy-today.com/Scharf.html
STM images
      http://www.di.com/Theater/Main.html
      http://www.almaden.ibm.com:80~/vis/stm/gallery.html
Micrographs as art
      http://cammer.home.mindspring.com/art/smkmell/smkmell.htm
      http://www.lucent.com/microscapes/microscapes.html
Animations
      http://research.amnh.org/idl/scivizgallery/galleries.html
Image quizzes
      www.exn.ca/@discovery.ca (select @Discovery.CA/Small Wonder)
      http://www.theimage.com/closeup/closeup.html
Microscopy experiments
      http://www.byu.edu/acd1/ed/microscopy/
      http://www.mos.org/sln/SEM/resources.html

MICRO lessons
> http://www.ccmr.cornell.edu/microworld

Microbiology lessons
> http://www.asmusa.org/edusrc/edu39.htm

Amateur microscopy
> http://www.microscopy-uk.org.uk
> http://seansys.tierranet.com/AmMicSci/amswr.mv?next+915904955

Histology links
> http://www.histology.to/links.html#anchor201911

Histology atlas
> http://www.udel.edu/Biology/Wags/histopage/histopage.htm

Microbes
> http://www.nhm.org.microbes

Microbe zoo
> http://commtechlab.msu.edu/ctlprojects/dlc-me/

Microscopy of food
> http://www.cyberus.ca/~scimat/

Crystals
> http://www.crystal-land.com
> http://www.pbrc.hawaii.edu/~kunkel/
> www.cellsalive.com

Snowflakes
> http://www.mee-inc.com

Diatoms
> http://www.BGSU.edu/departments/biology/algae/index.html

Dinoflagellates
> http://www.comet.net/gek/phytob.htm

Living cells
> http://www.cellsalive.com/

Living mites
> http://www.feic.com/esem/index.htm

3D Images
> http://www.microscopy-uk.org.uk/amateurs/mic3d/3dfront.html
> http://www.feic.com/gallery/gallery.htm

Demo of refraction
> http://covis2.atmos.uiuc.edu/guide/optics/ html/refr-effect.html

Virtual light microscope
> http://www.microscopy-uk.org.uk/prodir/software/softmol.html

Virtual SEM
> http://micro.magnet.fsu.edu/primer/java/electronmicroscopy/magnify1/index.html

Powers of 10 chart
> http://mse.mcmaster.ca/research/micro/

Microscope history
> http://www.sciences.demon.co.uk/whistmic.htm
> http://www.utmem.edu/personal/thjones/hist/hist_mic.htm

Leeuwenhoek microscope
> http://www.sirius.com/~alshinn/

Home-made microscope
> http://www.mos.org/sln/sem/myomicro.html

Buying a microscope
> www.msa.microscopy.com/ProjectMicro/BuyMicroscopes.html
> http://www.diwalk.demon.co.uk/novice/choice.htm

Assessment Suggestions

Selected Student Outcomes

1. Students improve in their ability to make detailed observations of a variety of objects and organisms and gain insight into the central role of careful observation in scientific inquiry.

2. Students increase their understanding about why certain objects magnify and how the curve of a lens (or water drop) affects its magnitude of magnification.

3. Students improve in their abilities to use hand lenses and compound and/or dissecting microscopes and to accurately interpret what they see.

4. Students gain experience with the scale and structure of matter, and the diversity of the natural world, contributing to later understandings about molecular biology, atomic structure, and many other fields of science.

5. Students gain insight into the role of technology in science and how a microscope enables people to perceive objects, phenomena, attributes, and characteristics that are otherwise unobservable. They also gain initial insight into the ways that magnifiers, microscopes, and related technologies have been and continue to be instrumental in advancing human knowledge in many scientific disciplines and fields of inquiry.

Built-In Assessment Activities

In Activity 1, Up Close, students explore the magnifying properties of a variety of objects and are asked to explain what makes a good magnifier. The teacher can observe students as they work at the station and can analyze responses in the student observation booklet. If there are opportunities for discussion, the teacher can use questioning strategies as needed to help students realize that the amount of magnification relates to the amount of curvature. Student oral respones in the prior whole-class lesson on magnifiers and microscopes can provide information on what students know as they begin the unit. (Outcomes 1, 2, 5)

At most stations, including Dots and Dollars, Fabrics, Salts, Sand, Small Creatures, Brine Shrimp, and Pond Life, students are asked to first observe through a microscope and/or hand lens, then to carefully draw and describe what they see. The teacher can observe the students as they perform these tasks and can look for growth in student work over the course of their writing and drawing in the student observation booklet. Detailed student drawings can provide a very rich assessment resource for the teacher, and may allow some students who may not excel in writing to nevertheless demonstrate a high level of conceptual understanding and observation skills. (Outcomes 1, 3)

At the Salts, Sand, and Kitchen Powders stations, students investigate matter and the differing forms that small mineral and chemical substances take when observed through instruments that enhance human vision. Teachers can determine from student work at the station and in the booklet how accurately students are able to interpret what they see, and, in the case of Salts and Kitchen Powders, whether students are able to use what they have observed and learned to solve several "mysteries." (Outcomes 3, 4, 5)

In the Small Creatures, Brine Shrimp, and Pond Life stations, students gain an enhanced appreciation for the diversity of life and the complex nature of organisms that we can only see clearly in detail through use of a hand lens or microscope. Teachers can analyze student descriptions and drawings in the booklet to assess growth in ability to describe organisms in detail. This work may also reflect improvement in student ability to use hand lenses and microscopes effectively, especially when the subject of observation is in motion. (Outcomes 1, 3, 4, 5)

In the Fingerprint Ridges station, students are asked to analyze their own fingerprints by ridge attributes and then to apply what they've learned to compare the ridge details in two other prints. Teachers can observe students as they work on these tasks, and look for abilities related to detailed observation, persistence, and application to a new situation. Direct observation and the student booklet can also assist teachers in determining how well students are able to adapt to the main tasks at the station, where they are asked to conceive of fingerprints in a new way— not in the usual classification system, but by detailed observation of ridges—a method often used in the real world by forensic technicians. (Outcomes 1, 2, 5)

In the Fabrics station, students are asked to infer—from their observations with their eyes, a hand lens, and a microscope—how a piece of fabric was made. They are asked to explain why they think the fabric was woven, or knitted, or pressed. The teacher can assess their responses for attention to detail, logical thinking, and ability to visualize and explain how structure may relate to manufacture. (Outcomes 1, 3, 4)

In the Dots and Dollars station, students gain insight into the technological processes of printing and the creation of color. Through the booklet and class discussions, teachers can gain insight into student observation abilities and their initial ideas about how color is created. Depending on student level and experience, teachers can encourage students to explain, based on their observations, how they think color printing happens. Through such a discussion, teachers can see how willing students are to venture possible explanations, and how logically they approach the task. (Outcomes 1, 3, 5)

Additional Assessment Ideas

Students can prepare a full investigation based on a topic suggested by their explorations at one or more of the stations. Depending on the grade level and prior experience, they may need guidance and additional activities to assist them in planning and structuring such scientific investigations. The teacher can assess the depth of content investigated by the students, as well as their capabilities in and understandings of scientific methods of inquiry. (Outcomes 1, 5)

Students could be asked to write a story on their adventures in a microscopic world, such as the world they have observed in pond water, a world of crystalline shapes, or any other environment suggested by what they have seen at the learning stations. The teacher should provide some parameters and expectations for the story, such as telling students that they should exercise their creative imagination, but they also are to: (1) include detailed descriptions of at least three of the organisms or substances they observed; (2) discuss how a microscope works; and (3) give a sense of the relative sizes of the parts of the environment they describe. These are just examples—you may want to "focus" on other learning goals in your assignment of the story. (Outcomes 1, 4, 5, depending on specifics of assignment)

Many of the "Going Further" activities suggested in connection with the *Microscopic Explorations* stations can be used as assessment activities. For example:

  • For the Fabrics station, students make pipe cleaner models of woven and knitted fabrics, providing an excellent (and three-dimensional) assessment of how well they understand and can represent how fabrics are made. (Outcome 4)

  • After the Kitchen Powders station, students create additional "mystery mixtures" and challenge other students to identify the substances in the mixture. (Outcome 4)

  • A "Going Further" suggestion after the Pond Life station asks students to make a detailed comparison between brine shrimp and *Daphnia*, allowing the teacher to gauge student abilities to observe, record, and draw conclusions, as well as assess growth in student understanding of and appreciation for the natural world. (Outcome 4)

  • Presentation of the GEMS unit *Color Analyzers* in connection with *Microscopic Explorations*, especially the Dots and Dollars station, can provide additional assessment information on student comprehension of physical science concepts related to light and color. (Outcome 4)

  • Assessment activities built into the GEMS unit *More Than Magnifiers* assist teachers in determining how well students understand that lenses have properties that can be measured, that curvature is a predictor of magnifying power, that some lenses are better for certain purposes than others, and that—with just two lenses—many optical instruments can be made. (Outcome 2)

All of the GEMS units listed as possible "Going Further" activities for the different stations contain their own assessment components. *Insights and Outcomes: Assessments for Great Explorations in Math and Science* (also known as the GEMS assessment handbook) also provides a number of detailed case studies with student work to help teachers better understand and implement modern approaches to performance-based mathematics and science assessment.

Literature Connections

Arrowsmith
by Sinclair Lewis
P.F. Collier & Son, New York. 1925   Grades 7–Adult
    This famous novel, by the noted American writer, although
dated, brings the excitement of scientific inquiry to life while exploring
the intersection of science and society, as well as examining issues of
scientific integrity and ethics. Dr. Paul H. DeKruif, the scientist-author of
Microbe Hunters, collaborated with Lewis.

Color
by Ruth Heller
Putnam & Grosset, New York. 1995   Grades 1–Adult
    This book explains how artwork gets reproduced from pictures
to the printed page. With a four-page acetate printer's proof bound right
into the book and striking illustrations, this unique, interactive picture
book in verse will appeal to many readers. It connects well to the Dots
and Dollars station.

The Disease Fighters: The Nobel Prize in Medicine
by Nathan Aaseng
Lerner Publications, Minneapolis. 1987   Grades 6–12
    Many major medical discoveries are described, including the
cause of malaria and the cure for tuberculosis. Includes much basic
information on how microbiological breakthroughs have contributed to
the battle against disease.

The Electron Microscope: A Tool of Discovery
by Aaron E. Klein
McGraw-Hill, New York. 1974   Grades 7–12
    This book provides basic information on the electron microscope,
tracing its development, describing how it works, and highlighting the
major discoveries made with it.

Frontiers of Medicine
by Lionel Bender
Gloucester Press, New York. 1991   Grades 4–8
    This 32-page book, part of the Through the Microscope series,
examines the advances in medical science made possible by the use of
microscopes, from early studies of plants and insects to the eventual
discovery of bacteria.

Greg's Microscope
by Millicent E. Selsam; illustrated by Arnold Lobel
HarperCollins, New York. 1963   Grades K–3
    Greg gets a microscope and has a fascinating time looking at
sugar and salt crystals, thread, hairs, and cells. Greg's observations of
these common items occurs through a good discovery approach. Unfortunately, gender roles in Greg's family are depicted in a stereotypical
manner, which diminishes the appeal of the book.

Faith is a fine invention
For gentlemen to see;
But microscopes are prudent
In an emergency.
        Emily Dickinson
        Poems, 1880

How Did We Find Out About Germs?
by Isaac Asimov; illustrated by David Wool
Walker, New York. 1974   Grades 4–9
From Asimov's How Did We Find Out... series, this book traces the development of human knowledge about germs from their first sighting under the earliest microscopes to the more sophisticated 20th century technologies and modes of attack on disease-causing agents.

Medical Technology: Inventing the Instruments
by Robert Mulcahy
Oliver Press, Minneapolis. 1997   Grades 7–12
Profiles the life and work of seven scientists who made major contributions to important medical inventions, including Leeuwenhoek and the microscope, as well as those associated with the thermometer, stethoscope, X-rays, radiation, anesthetics, and the electrocardiograph.

Microbe Hunters
by Paul DeKruif
Harcourt, Brace & World, New York. 1954   Grades 8– Adult
This classic book includes exciting narratives documenting the scientific discovery of microbes and how such discoveries led to new understanding and prevention of deadly diseases.

The Microscope
by Maxine Kumin; illustrated by Arnold Lobel
HarperCollins, New York. 1984   Grades 4–Adult
Relates in rhyme the story of the famous Dutch scientist Anton van Leeuwenhoek who, though a shop owner, ground lenses for use in his own microscope. The illustrations depict many of Leeuwenhoek's remarkable observations.

What's Smaller Than a Pygmy Shrew?
by Robert E. Wells
Albert Whitman, Morton Grove, Illinois. 1995   Grades 1–6
A pygmy shrew is among the tiniest of mammals. A ladybug is even smaller. In this book children can find small things not ordinarily seen.

Zoom
by Istvan Banyai
Viking, New York. 1995   Grades 2–Adult
This wordless picture book presents a series of scenes, each one from farther away, showing, for example, a girl playing with toys which is actually a picture on a magazine cover, which is part of a sign on a bus, and so on.

Sample Letter to Parents

Dear Parents,

Your child will soon get to participate in a science unit called *Microscopic Explorations*, developed by the Great Explorations in Math and Science (GEMS) program at the Lawrence Hall of Science, in cooperation with the Microscopy Society of America. Through this unit, students learn to use a microscope and in so doing are introduced to the many fields of science opened up by this marvelous instrument. There are numerous items needed for this series of activities, so I am asking for your help. Below is a partial list. Please see if you have any of the following items, or if you know of others who do. Contact me first, so we make sure not to duplicate our efforts! We will need the materials by _____. It will be appreciated if all your donations have been cleaned. Thank you very much for your help!

Sincerely,

____ 24 small ziplock bags (qt. size)

____ 24 manila folders (letter size)

____ 1 container rubber cement

____ 40–60 hand lenses

____ up to 100 clear plastic deli-lids

____ extension cords

____ power strips

____ table lamps

____ duct tape

____ several empty clear vials, pill bottles, or water bottles with lids, any size

____ several clear, flat-sided plastic or glass containers, any size

____ 2 two-liter clear plastic bottles (not tinted)

____ 8–10 clear, flexible plastic cups (about 6–10 oz. size)

____ 12 rolls of clear transparent tape with dispensers (not the Magic™ type)

____ 5 rolls of Magic™ tape, ¾" wide, with dispensers

____ 10 medicine droppers

____ 12 plastic depression slides

____ about 10 clear, flattened marbles

____ about 10 clear, round marbles

____ assorted fabric samples (woven, knitted, pressed)

____ Velcro, a few short pieces of both sides

____ 2 packages Epsom salt

____ 2 empty deli or cottage cheese containers with lids

____ a few yogurt containers or film canisters

____ 1 packet brine shrimp eggs

____ a sprig of *Elodea*, an aquarium plant

____ lots of pipe cleaners

____ 50–60 styrofoam peanuts

____ 20–30 popsicle sticks

____ 20–30 straws

____ 20–30 toothpicks

____ 1 film negative

____ a few black and white photos

____ a few new color photos

____ a few business cards

____ Georges Seurat pictures

____ a few magazine pictures

____ spoonfuls of sand from different places, labeled with place gathered

____ a few rock samples

____ a few shells

____ a few coral samples

____ 2 small squeeze bottles of liquid dishwashing detergent

____ 1 hot pot

____ 1 metal spoon

____ 2 plastic spoons

____ a few dead insects

____ live brine shrimp adults

____ pond water with living organisms

Volunteer Sheet, SIDE TWO

Your goal is to assist the students in making their own discoveries, while keeping the activity safe and the mess under control. Read the sign at the station and try the activity yourself so you know what the students will be investigating. Help students follow the procedure by asking a question or presenting a challenge, rather than by demonstrating how it is done. Save demonstrations as a last resort for situations when the students are not successful on their own, even with coaching.

The goal of the activity is to improve observation skills, improve student abilities in using the microscope and magnifying lens, and to encourge students to learn more about the science introduced by the different stations. Resist the temptation to give explanations to students. Ask open-ended questions, such as:

- What have you discovered?

- What does it look like to you?

You may also want to provide further challenges, such as:

- What else would you like to find out?

- How might you do that?

Encourage students to share their discoveries with one another at the station.

If students appear to be out of control, using materials in a way that seems unrelated to the station activities, or treating the microscopes in a way that might cause damage, please steer them back on task. Be sure to intervene if you see an unsafe behavior. Keep in mind, however, that what may sometimes appear to be "fooling around" may lead to some of the deepest learning experiences. Some of the greatest scientific discoveries have been made when scientists freely explored materials or looked at a problem in a new and creative way!

150

Microscopes
Tips for Safety and Success

• Do not pick up or tilt the microscopes. If you ever do need to pick up a microscope, always use two hands, and keep it upright (otherwise the eyepiece will fall out of the top of the microscope).

• Be sure to keep the microscope dry. If you spill water, dry it immediately. Always be sure the bottom of your slide is dry before putting it on the microscope stage.

• Having trouble using the microscope? Here are some things to check:

Is the shortest objective "clicked" into place?
Change to the shortest objective by turning the ring it's attached to. You'll feel a little click when the objective is lined up with the hole in the stage.

Did you focus?
Turn the focus knob until the object looks clear to you. On compound microscopes there are two sets of focus knobs. Use the bigger knobs first (coarse focus), then use the smaller knobs (fine focus).

Is the light adjusted?
Set a table lamp or fluorescent tube so the light is shining on both the stage and the mirror. Tilt the mirror so it shines the light up through the stage. Check the diaphragm (the rotating disk with different sized holes) under the stage to make sure a large hole is lined up with the hole in the stage.

What if you see your own eyelashes?
Look up at something far away, and then look through the microscope again with the same focus.

Does everything look backwards and upside down?
Compound microscopes "flip" the image! You will get used to this when you're studying something with a compound microscope. Dissecting microscopes do show the image right-side-up.

Microscopic Explorations

Student Observation Booklet

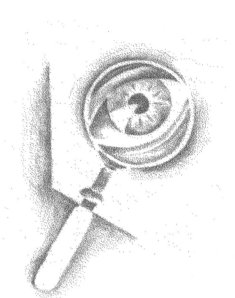

Name _____

Festival Closing Questions

What did you find out from doing these activities?

_____
_____
_____

Did you see something that you have never seen before?

_____
_____
_____

What was your favorite thing in the festival?

_____
_____

What would you still like to find out?

_____
_____

10. Pond Life

A. Draw and describe one plant and one animal.

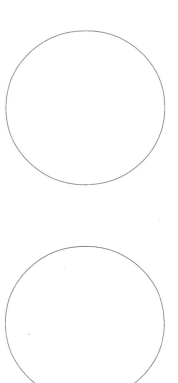

Plant   **Animal**

_____

_____

B. What colors and shapes can you see on the plant? What other details do you see on the plant?

_____

_____

C. What parts of the animal do you see? Legs? Wings? Eyes? Antennae?

_____

_____

D. How is the water animal moving?

_____

_____

1. Up Close

A. Which objects make things look bigger?

_____

_____

_____

B. How are the objects that magnify similar?

_____

_____

_____

C. What makes a good magnifier?

_____

_____

_____

2. Fingerprint Ridges

A. Tape your fingerprint in the box. Do you have any ridges like the ones below? Draw lines to label some of your ridges.

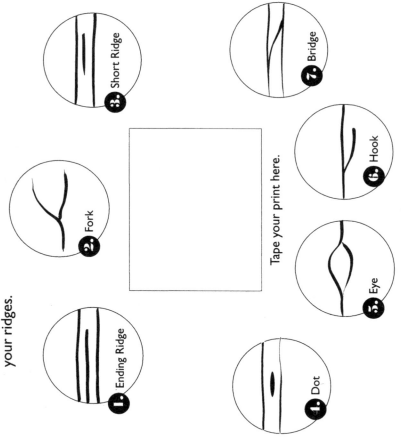

1. Ending Ridge
2. Fork
3. Short Ridge
4. Dot
5. Eye
6. Hook
7. Bridge

Tape your print here.

B. Compare the ridge details in these two fingerprints. Circle any differences you see.

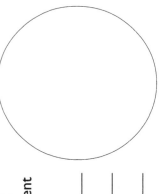

Print A   **Print B**

---

9. Brine Shrimp

A. Draw and describe an adult brine shrimp. How does it move?

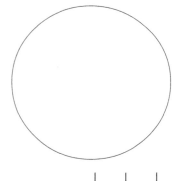

_____
_____
_____
_____

B. Is the brine shrimp adult a male or a female? How do you know?

_____
_____

C. If you see a larva, draw and describe it. How is it different from the adult?

_____
_____
_____
_____

D. How many brine shrimp eggs do you think are on the slide? Write your estimate on the paper at the station.

8. Small Creatures

A. Pick one of the creatures to study. Draw what it looks like through a hand lens.

B. How many legs does it have? What do the legs look like?

_____

_____

C. Does it have wings? What do they look like?

_____

_____

D. What do the eyes look like? The mouth?

_____

_____

E. What other body parts do you see? Describe them.

_____

_____

F. Using the microscope, draw **one body part** of the creature in detail.

---

3. Dots and Dollars

A. What do you notice when you look at all the items with a hand lens?

_____

_____

B. What do you notice when you look at the items with a microscope?

_____

_____

C. When looking at the colored newspaper:

• What do you notice about the dots?

_____

_____

• Are all the dots the same size? Are they the same distance apart?

_____

_____

• How many different colors of dots can you find? What colors are they?

_____

_____

4. Fabrics

A. Pick one fabric. Draw and describe it when you look:

• with just your eyes

• with a hand lens

• with a microscope

B. Do you think your fabric is woven, knitted, or pressed? Why do you think so?

_____

_____

_____

7. Kitchen Powders

A. Pick any two powders. Carefully draw and describe them.

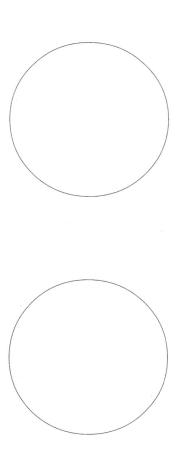

Name of Powder _____   Name of Powder _____

B. What is the mystery powder? How do you know?

_____

_____

C. What is the mystery mixture? How do you know?

_____

_____